The
Sweetness
of
Venus

The
Sweetness
of
Venus

A History of the Clitoris

SARAH CHADWICK

For further information, contact:
Wild Pansy Press
825 Wildlife
Estes Park, CO 80517
ISBN: 978-1-7362988-3-1

*This book is for all those who have a clitoris,
or the opportunity to engage with one.
With love.*

Contents

Introduction

The Hippopotamus in the Room

When my children were young, they would take their evening bath together. The oldest, a boy, would sit with his back to the smooth, curved end of the bath, while my second son sat with his back to the faucet. My daughter, the third child, would sit in the middle, facing into the bathroom—a bath-time position hierarchy established by siblings and as entrenched as which seat you get in the car. (How many families do you know where the youngest child is always made to sit in the middle of the back seat?) For as long as I can remember, the boys sometimes played with their penises, trying to draw attention to them and being ignored. They would soon move on to other bath-time play, like covering their faces with bubbles, which is why I wasn't perturbed by their behavior. It wasn't as if they were doing this in the school playground or at teatime.

During one such bath-time, when my daughter was three and a half, she stood up, knees akimbo, her thumb and forefinger pulling on her clitoris, shouting, "I've got one! I've got a willy! Look, I've got a willy too!" Complicated layers of simultaneous thought occurred to me in the seconds following her announcement. Question to self: what is the appropriate response? Interest: who knew she felt left out of the willy parade? Academic line of enquiry: is this Freud's penis envy? Panic: what are we going to call her clitoris? Feminist self: this moment matters. Warning: don't give her shame.

"Yes, darling, you have," I said. "It's called a *clitoris,* and it is every bit as good as a willy and will give you great pleasure." At which point she sat down and poured more bathwater tea with a plastic teapot that had migrated from the toy kitchen.

The following morning, while walking with her brothers to school, my tousle-haired daughter is trotting along the pavement, behind her brothers but ahead of me, when she goes up to a stranger sitting on a low wall. Next thing, I hear her say enthusiastically, "Do you know, I've got a . . ." At this point I know what is coming and wonder whether I've handled the bath-time discovery quite so well after all. I'm calculating how noisily one would have to sing to drown out her words, when she says, "I've got a hippopotamus between my legs." How she got from *clitoris* to *hippopotamus* I don't know, but I have never been more grateful for a malapropism. Maybe she'd heard the hard, consonant sound in the middle and the *s* at the end, and in searching for the term a day later she wanted a word that captured the vastness of her news and the rarity of her discovery.

Why was I so embarrassed that she might say "clitoris"? Why isn't there a friendly slang term for the clitoris in English that I could have used instead, and does it matter that there isn't? Why don't we tell girls and boys about this aspect of a woman's anatomy? I was led to revisit this moment when my daughter was much older and we began to have conversations about sex, the orgasm gap that seems to exist in some hetero-sexual encounters, and the disparity in levels of information about the clitoris and female sexual pleasure. How did female sexual pleasure and the clitoris get written out of the script? How come there is such a difference between attitudes to and expectations of male sexuality and female sexuality? What cultural threads have shaped the taboo that seems to exist against the clitoris? How important is it to vulva owners in terms of sexual pleasure? How informed do people feel? I found myself researching the history of the clitoris in the worlds of anatomy and science and asking what roles religion, philosophy, politics, and psychology have played in determining attitudes toward female sexuality and the clitoris. I also became interested in how these areas of human thought have influenced our culture today—in terms of language, the written word, and visual representation.

This book unravels the fascinating and troubled history of the clitoris, from her discovery to the present day. It is not a how-to manual, and there are no photographic images. It is a surprising, funny, straight-talking historical narrative about the clitoris in Western culture. It is a story of denial, marginalization, brutality, and lies. Like America, much of the clitoris's story has been defined by those who discovered her. The true extent of the clitoris was only fully mapped with 3D imaging in 2005, and this knowledge is game-changing. Why did it take so long? This book will tell you and give you the arguments to challenge history. Now that we know the story of the clitoris, it's time to tell it.

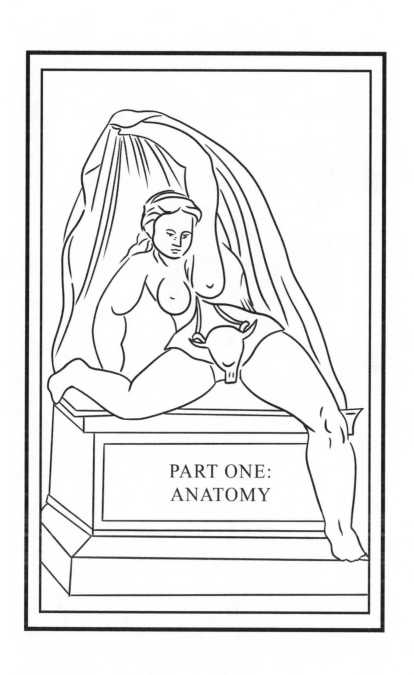

PART ONE:
ANATOMY

1
Everyone Must Have a Penis!
Obviously, right?

There is an ancient Indian fable about six blind men who encounter an elephant. As each man touches a different part of the animal, he announces his discovery.

The first blind man puts out his hand and touches the elephant's side. "How smooth! An elephant is like a wall." The second blind man puts out his hand and touches the trunk. "How round! An elephant is like a snake." The third blind man puts out his hand and touches a tusk. "How sharp! An elephant is like a spear." The fourth blind man puts out his hand and touches a leg. "How tall! An elephant is like a tree." The fifth blind man puts out his hand and touches an ear. "How wide! An elephant is like a fan." The sixth blind man puts out his hand and touches the tail of the elephant. "How thin! An elephant is like a rope."

Initially, I imagined creating a timeline to explain people's awareness of the clitoris, like those you find in science textbooks and encyclopedias, charting a progression from ignorance to enlightenment in terms of the scientific knowledge of its anatomy. This proved impossible because ideas and beliefs about the clitoris have been patchy and inconsistent. There isn't a straight line of thinking that can be plotted. For many centuries, conflicting ideas coexisted: some anatomists didn't think the clitoris was real, while other anatomists believed it was key to female sexual pleasure. However, as I kept researching, it became apparent that there was one androcentric (male-focused) thread holding much of the disparate thinking together: the belief that woman is a version of man. This was one thing everyone agreed on.

Androcentric thinking about women's bodies led to two theories: either the vagina was an inverted penis, or the clitoris was a penis equivalent. Either way, everyone must have a penis, and the male version was the superior model. This idea can be seen throughout the history of Western thought about women's anatomy, beginning with the Greeks and Romans. The Alexandrian anatomist and dissectionist Herophilus claimed as early as the third century BCE that women had testes with seminal ducts on each side of their uteruses.[1]

In the vagina-is-an-inverted-penis version of a woman's body, erroneous and damaging notions are perpetuated: the beliefs that women experience sexual pleasure the way men do and (this one gained real momentum at the end of the 19th and through the 20th century) that women were inadequate if they didn't experience an orgasm vaginally. Mutually satisfying heterosexual orgasmic delight for the masses was never given a fair chance under this model. The clitoris was sidelined and not recognized as related to sexual pleasure within any "normal" framework of sexual engagement.

Alternatively, in this world where all things were measured against men, the clitoris was sometimes regarded as the equivalent of the penis. This perception of the body acknowledged that people have sexual pleasure as a result of the clitoris. Medieval, early modern, and Renaissance thinkers who advanced this view, often joyously, tended to be overshadowed by the vagina-as-penis champions, and it wasn't until the end of the 18th century that this model began to seriously vie for scientific attention.

That's when the trouble really began. The clitoris's very existence challenged the idea that the penis was the provider of all things good and opened up the possibility of a parallel universe in which women might eschew coital sex and take control of their own physical pleasure. God forbid—what would happen then? The clitoris unleashed a threat to an establishment that went into overdrive to defend itself, insisting that the clitoris was a substandard substitute for a penis; as such its inadequacy was seen to reflect other inadequacies of women. Women were just small, poor versions of men, lacking in moral judgement, reason, physical strength, *and* a decent penis. Oh, and yikes, this one was really problematic: women could (wait for it . . .) indulge in masturbation. Obviously, given women's lack of self-control, this possibility was reason for civic concern. While masturbation by either sex had always been frowned upon, it now became a particular female no-no. Serious men in the 18th and 19th centuries invested a lot of time and energy worrying about women literally taking matters into their own hands.

You can have a lot of fun thinking about human bodies in gynocentric (women-focused) terms. Inverting the narrative highlights the absurdity of the historic insistence that woman is a version of man. It also forces us to recognize how errone-ous ideas can become embedded in a world view and end up difficult to challenge.

In a gynocentric version of the world, the penis becomes a poor, clumsy, cumbersome version of a clitoris, an early Nokia mobile phone, say, to the slim, multifunctional, sensitive touch-screen phones of today. Men are seen as the ones lacking, not women. Men don't have a vagina, a womb, or breasts, and their testicles are ridiculed as a substandard design, inefficient due to their overproduction of sperm and vulnerability to being hurt. These failures of a man's body are symptomatic of their deficiency in other areas, such as a masculine failure to provide nurturing care in politics and a wanton predisposition to waste natural resources. But this isn't how the narrative has gone—instead, it is rigidly driven by the concept that man is perfect and woman, measured against him, is imperfect. Either way, the thinking is binary.

The distortion of women's anatomy that has become embedded in European and American culture has impacted far more than attitudes and beliefs relating to sexuality. It has affected design and functioning in car-crash safety, testing dummies built to men's proportions; bullet-protection vests that don't accommodate breasts; hip-replacement joints sized for small, medium, and large men only; voice-recognition technology that is less effective when sampling women's voices; step counters calibrated to man-size steps; the dimensions of piano keyboards and wrenches; and the setting of an ideal office temperature.[2] Women have spent their lives accommodating themselves to a man's world. Cue James Brown.

A recent article for *The Guardian*[3] opened with a story told by one of my favorite U.K. broadcasters, Sandi Toksvig. When Toksvig was studying anthropology at university, one of her female professors held up the image of an antler bone with 28 markings on it. "This," said the professor, "is alleged to be man's first attempt at a calendar. Tell me, what man needs to know when 28 days have passed? I suspect that this is *woman's* first attempt at a calendar."

It's a joke. We all know that the words *man* and *mankind* are frequently used to refer to humans and humankind, but that's the point. When Apple launched its comprehensive health tracker in 2014, it could monitor blood pressure, steps taken, blood alcohol level—even molybdenum and copper intake—but not menstruation. It was designed with mankind in mind, not humankind.

Over the centuries, it did not occur to anyone in a position of authority that women might either be different from men or be lucky enough to have two penises: that would have been viewed as bonkers, shocking, and probably sacrilegious. No, it was one or the other: a woman's equivalent of the penis was either the vagina or the clitoris. Not only was neither theory accurate, but they were both destructive. They led to interventions for women such as third-party stimulation provided by medics, clitoral cleaning and surgery, and—if you were lucky and got the least bad option—hours in a therapist's chair being talked out of any engagement with your clitoris at all.

How did women come to be seen as anatomical versions of men?

Imagine you're relocating and looking for a new gynecologist. Scanning the biographies of the doctors in your area, you come across one that reads:

> *Claudius Galen:* Born 130 CE. After working as surgeon to the gladiators, Galen continued his training in the field with the Roman army, before becoming doctor to Emperor Marcus Aurelius. Special interests: autopsies on Barbary apes, pigs, and dogs. A prolific, widely published, and influential writer in the fields of medical science and philosophy, Galen is greatly influenced by Aristotle, who believed that "the male is by nature superior and the female inferior."[4]

What are the chances you'd think, "That's my guy! One hundred percent I need to get on his list!"? You wouldn't. But for centuries he became the preeminent source of knowledge about female anatomy and was the go-to for many European thinkers and anatomists.

> *Question:* What was Galen's revered, scientific knowledge about woman's sexual anatomy?
> *Answer:* The vagina is an inverted penis. What clitoris?

Galen maintained that the difference between the sexes is seen in the development of the reproductive organs. He drew analogies between the ovaries and the testicles, and the uterus and the scrotum, as Herophilus had done before him. Galen said that essentially men and women had the same organs, but that those of women remained internally located due to a lack of heat. "Just as the human species is the most perfect of all the animals, within the human the man is more perfect than the woman, and the reason for his perfection is his greater heat, for heat is the first instrument of nature."[5]

You can add "heat" to your list of things that women have been told they lack. The absolute perfection of man was a given; it was an opinion stated as fact so frequently and authoritatively that it wasn't questioned. Thus began the notion of a unisex[6] body, with a woman's body framed as that of a failed man. Reading Galen's instructions on how this inside-out anatomy works is like trying to follow an origami booklet:

> Turn outward the woman's, turn inward so to speak, and fold double the man's, and you will find the same in both in every respect . . . Think first please, of the man's [external genitalia] turned in and extending inward between the rectum and the bladder. If this should happen, the scrotum would necessarily take the place of the uterus with the testes lying outside, next to it on either side ...Think too, please, of the converse, the uterus turned outward and projecting.

> Would not the testes [ovaries] then necessarily be
> inside it? Would it not contain them like a scrotum?
> Would not the neck [the vagina and cervix], hitherto
> concealed inside the perineum but now pendant, be
> made into the male member? You could not find a
> single male part left over that had not simply changed
> its position.[7]

Put another way, the penis is an outie vagina, and the head of
the cervix equates to the head of a penis.

There are three failures in this line of thinking. First, it
entirely misses the point about the abilities of the cervix and
vagina to stretch to allow birth. The penis stretches too, but
I don't think Galen empirically tested the stretchiness of the
penis with his gladiators, do you? Just think of the fuss men
make passing a kidney stone. The penis expands and contracts
in an entirely different way to that of a vagina and is alive with
nerves and feeling. Lucky then that the vagina is not—or how
would birth happen? Clearly the cervix is not like the head of a
penis at all. The famous sexologist Alfred Kinsey, in his study
Sexual Behavior in the Human Female, reported on the topic of
vaginal feeling that "among women who were tested in our
gynecological sample, less than 14 percent were at all conscious
that they had been touched."[8] Which makes sense when you
think about it—otherwise Tampax would be the brand name
of a sex toy.

Second, if the vagina-as-penis model was true, then by
correlation, the thrusting stimulation that brings a man to
orgasm should also stimulate a woman to orgasm, and within
a similar time frame. If only. There is a neat yin–yang to this
theory, and a romance about the design of two bodies fitting
together so neatly in coital and reproductive satisfaction. It's
easy to see how this would become the dominant narrative
as to how sex works; however, it is not a perfect narrative arc,

since intercourse alone is not orgasmically effective for many women. In fact, studies consistently show that it is not orgasmic for *most* women.

Yet this vision of sexual mutuality lasted well into the 20th century and continues today. For example, in the chapter on "Having Sex" in one of the best-selling sex-education books for teenage boys,[9] readers are told that as the penis slides back and forth in the woman's vagina it "stimulates the nerves at the end of the penis and in the vagina and clitoris." But the vagina has only a small number of nerves, and the clitoris is not stimulated by this action in *all* women. It would be more accurate to say that it "stimulates the clitoris in a quarter of women." The comment that these sensations "often result in orgasm" should be rewritten to say, "These sensations often result in orgasm for men. For women, more direct clitoral stimulation is often needed to experience orgasm." There are lovely men who long to bring their women partners to climax and believe that with enough thrusting they can get there. When the starburst doesn't happen, they feel inadequate. Maybe they didn't go long enough? Maybe their penis is too small? If only they knew about the clitoris in all her glory. If only women had the courage to tell them.

Which brings us to the third and most pressing issue with Galen's science. In this esteemed scientific model, where woman is a version of man and the vagina is the penis counterpart, all the male genital anatomy can be accounted for but the clitoris is like that random bolt left over after you've constructed a set of IKEA shelves. It doesn't have a more perfect, male, anatomical twin. As a result, the clitoris is either a fleshy aberration or goes entirely unnoticed.

Galen's biological views concerning women were not seriously challenged until the 17th century. He died in 200 CE, but his work was referenced and re-referenced in medieval literature on the human body—and during the Renaissance, some 1,300 years after his death, it was frequently translated from the original Greek into Latin for European physicians.[10] Galen's texts on anatomy and medicine were not published as oddities, but as the serious, erudite work of a clinician, master anatomist, and diagnostician. And his writing heavily influenced Renaissance thinkers.

Although Andreas Vesalius, medical authority of his day, took issue with the fact that much of Galen's biological science had been drawn from animal dissections, he did not challenge the male benchmark or his vision of the anatomy of women; he just worked out what the origami looked like. Vesalius produced exquisitely detailed and artistic drawings of the human body that were copied throughout Europe, perpetuating the belief that women were inverted men. To the right is one of Vesalius's drawings from his much copied *On the Fabric of the Human Body in Seven Books,*[11] published in 1543. You might think that what you see looks suspiciously like a penis, but it is in fact a representation of a disembodied vagina. The testes at the top have become a womb.

This is not what you think it is.

Still finding it hard to visualize how Vesalius's penis-as-vagina works? Try this . . .

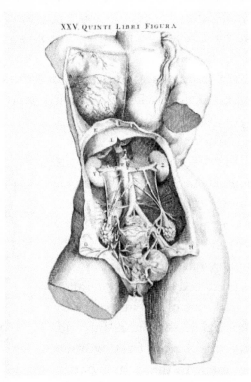

The perfect penis inversion.

Do you see the tip of the penis acting as a vulva at the entrance to the vagina? It is staggering how many copies of Vesalius's book survive today—more than 700 from the 1543 and 1555 editions[12] —and this indicates how widely it was distributed and valued. Versions of Vesalius's drawings were replicated in printed anatomy texts throughout Europe. For example, they were used by George Bartisch in *Kunstbuche* (1575); Valverde in *Anatomia* (1586); Vidus Vidius in *De Anatome Corporis Humani* (1594); and Jakob Henle in *Handbuch der Systematischen Anatomie des Menschen* (1866).[13] Even American surgeon and artist Frank Netter's *CIBA Collection of Medical Illustrations,* Volume 2 (1954), used the traditional format of side-by-side illustrations of the male and female genitals shown in situ, although Netter placed his illustrations over more fleshy renderings of male and female torsos rather than using the Renaissance trope of cutting away classical statues. Significantly in this edition, there was still no clitoris on the cross section that shows the penis and vagina side by side, although the tiny Skene's duct and Bartholin's gland make it. I am reminded of the old adage, "To a man with a hammer, everything looks like a nail."

The Vagina Is an Inside-Out Penis.
What's this little bit?

So, the dominant viewpoint throughout history deemed the vagina the counterpart of the penis. Under this model, the clitoris was not seen to belong to the realm of sexual pleasure at all, and extraordinary qualities were attributed to it.

It's an abnormal appendage.

One consequence of not understanding women's anatomy and the clitoris's role in sexual function was that it became an oddity, an appendage that influential people could argue was unique or even threatening.

Aristotle stated that people who were intersex were not an intermediate sex but had doubled genitals, with one set being stunted and inoperative.[14] A vagina indicated that you were a woman, but if the external part of your clitoris was too obvious, surgeons would declare that it was a penis, a poorly formed second genital organ. Aristotle's influence has reverberated through the centuries. Lanfranc of Milan's 1296 recommendation of what to do in these circumstances makes terrifying reading. Of those with intersex characteristics, he says:

> Some of these have one that is fully formed, the other not fully formed, and some have neither fully formed.

On the contrary, they have in the orifice of the vulva
some added flesh, which is sometimes soft, fleshy, or a
small and weak character, other times of a strong and
sinewy character. The fleshy piece is removed swiftly
with cutting instruments, and those parts left behind
with light cauterization; the natural flesh must always
be taken care of by means of iron, or through a ligature
with thread until all superfluity is taken off.[15]

While Lanfranc, revered by his disciples and academia
of the day, was brutally cavalier in his treatment of what was
believed to be a substandard penis (but was presumably a small
penis or a clitoris), he was adamant that if the appendage was
too penis-like, that if it was "truly hard and strong," then "in
no way touch that with the iron nor think to treat it with
anything."[16] The penis was sacred in a way that the misunderstood clitoris was not. Furthermore, he felt that any woman
with too much "muscular growth" in the clitoral area, or a
"superfluity of skin," would be deemed disgusting to men, and
he recommended that the "pannicle" should be amputated and
cauterized with gold until it was reduced to "the natural form."[17]

Today we know that 1% of the population are intersex.[18]
This means their reproductive or genital anatomy is considered
atypical because it doesn't fit the standard definition of belong-
ing to either a man or a woman, and their anatomy may be at
odds with their gametes (either reproductive sperm or reproductive eggs that they are born with). For example, someone
may have both vulval and testicular tissue. There are as many as
30 variations of intersex, and for these people external genitalia
does not determine the gender they will identify with as an
adult, or how they will express this gender. Nor does it determine what sex they are attracted to.[19]

Sadly, while researching this, I came across an account
from 2008 on the Intersex Society of North America website in

which the author recounts how her clitoris, which she describes as "that wonderful location of pleasure for which I had no name but to which I had grown quite attached," was surgically removed by the Children's Memorial Hospital in Chicago in 1985 when she was 12, because it was considered outsize.[20] Her account is not an isolated one but hopefully, finally, the world is waking up to the complexity of intersex and the need to be better informed and less proscriptive about what constitutes organs of sexual pleasure and how they should look, as well as the danger of prejudging how an infant will want to identify as an adult.

As an adjunct to this, the week after I wrote the above and following a prolonged campaign by the Intersex Justice Project, the Lurie Children's Hospital in Chicago issued a statement that included an apology to the people who have been harmed by intersex genital surgery and pledged to stop these cosmetic and medically unnecessary surgeries while they explored comprehensive and informed intersex care. Hooray.

People worry about the size of their clitorises, and this may not be surprising. Firstly, they have *no* idea what constitutes "normal" because they don't get to see many, and secondly, there has been such emphasis on looking right for the male gaze. I was shocked to learn that in Australia the rules on porn dictate that for some magazines the vulva should be airbrushed to "heal it to a single crease." We should be outraged by the use of the word "heal" here. It's not broken! It's not sick! What kind of message are we sending vulva owners and consumers about how they should look?

On the American Society of Plastic Surgeons website, under "Vaginal rejuvenation, surgical options" (apparently one of the fastest-growing trends in cosmetic-surgery procedures), clitoral-hood reduction is listed as an option for excess folds in this area, to "improve the balance in appearance of the female

genitalia." Vaginal rejuvenation also includes the so-called
Barbie vagina. I am a big advocate for surgery that improves
function—the vaginal-prolapse repair that I had some time
after my last baby is a joy, and I am thrilled on a daily basis that
I don't pee at the slightest jolt to my body—but I am saddened
that a body part as intimate and wonderful as the clitoris, or
any other part of the vulva, can cause shame and anxiety on
account of how it looks. The sheer fact that the surgery is called
vaginal rejuvenation and Barbie *vagina,* rather than Barbie
vulva, implies a lack of accurate dissemination of information
in this area of cosmetic surgery. *Vagina* has become a careless
but widely used byword for the vulva in our culture, and as I
explore later, this misnomer keeps broader discussions about
female genitalia at bay by focusing attention firmly on the
birth canal rather than on the external genitalia. However,
you'd think that experts would want to be accurate, to enable
an informed discussion, wouldn't you?

Confusion over the status of the clitoris—which, if on
the large size, was considered an insult to penises and a threat
to masculine sensitivities about femininity, or damaging to
the sexual confidence of women themselves—has not been
its only challenge. The 1486 *Malleus Maleficarum,* a popular
guide to finding witches, claimed that the clitoris was a "devil's
teat" through which Satan sucked out a woman's soul. It was
commonly believed that the devil hid his marks so that witches
could pass undetected in society as normal women. To flush
out these dangerous beings, witch-hunters were encouraged
to conduct full-body and genital searches in their quests for
evidence. The clitoris, the labia, and prominent moles were all
signs that a woman was in cahoots with the devil.

Between 1487 and 1520, twenty editions of the *Malleus
Maleficarum* were published, and another sixteen between 1574
and 1669. Its claims became established lore on both sides of

the Atlantic. In 1598, after Alice Samuel was hung for being a witch in England—in Warboys, Cambridgeshire—her jailer and his wife carried out a postmortem and documented "a little lump of flesh, in manner sticking out as if it had been a teat, to the length of half an inch." Initially they were reluctant to disclose it "because it was adjoining so secret a place which was not decent to be seen," but "not willing to conceal so strange a matter, and decently covering that privy place a little above which it grew, they made open show thereof unto diverse that stood by."[21] Proof indeed for all those who cared to look that the prosecution had been right in its course of action.

In New Haven, Connecticut, in 1653 Mary Staples reportedly tore off the clothes of her hanged neighbor, Goodwife Knapp, to show the vaunted witch-teats, bravely declaring, "Here are no more teats than I myself have, or any other women … if you but search your body."[22] This didn't deter those on a quest to find witches—and rather than acknowledge the truth, the audience tried to have Mary herself indicted. A clitoris was a dangerous thing, particularly if you were thought to consort with the devil and let him suck your soul from it.

The genital signs that indicated you were a witch are left out of most modern-day accounts of witch trials, probably because the narrative becomes too graphic, adding complicating sexual violence to already dramatic historical events. It would be unsuitable material for grade school. Shhh—we don't talk about what's between a woman's legs, even if she's a witch.

It's hard to imagine that being a cis woman and having a clitoris wasn't considered "normal," but most science remained ignorant of the reality of the clitoris for many years. Victorian times weren't much more enlightened, and it was believed that prostitutes, the morally lax, and women prone to masturbation could be identified by their singularly large ones. It was with surprise that Alexandre Parent du Châtelet, vice president of the French Health and Sanitation Department, reported in 1857 after studying Parisian prostitutes that the clitoris "was found to be of normal size in females of the most unbridled passions."[23] Don't get me started on the assumption here that the prostitutes chose their work because they were horny.

Similarly, Black women have historically had their sexuality erroneously defined by a colonial culture seeking to denigrate them through it. Early colonizers, because of an arrogant failure to understand any culture other than the European one they had come from, were quick to portray African women as promiscuous and hypersexual. European scientists were fascinated by what they believed to be the prominent labia and clitorises of their colonized subjects, and these observations became mythical evidence. "The lining of the body appears to be loose," wrote Dutch physician Olfert Dapper in 1668 of the genitalia of African women, "so that in certain places it dangles out."[24] Dapper never visited Africa, but relied upon the accounts of other travelers. Ethnographer Willem Ten Rhyne wrote that African women "have dactyliform appendages, always two in number, hanging down from their pudenda."[25] In these surveys of Black female genital anatomy, I wonder how many white women were seen naked as a control group? Few, I suspect. I am reminded of British artist Jamie McCartney's 2008 piece *The Great Wall of Vagina,* which is created from plaster casts of the vulvas of 400 women and highlights the incredible variety in shape and size. What is "normal"? It's definitely not the neat slit of the historical imagination or of *Playboy.*

The narrative about the sexual availability and promiscuity of Black women as evinced by their allegedly larger clitorises and labia suited a colonizing agenda of subjugation and rape, and Black women's sexuality was played out in all areas of Western culture, including the circus. In 1810 Saartjie (Sara) Baartman was trafficked into the U.K. from the Cape of Good

Hope to become an object of fascination as the "Hottentot Venus," first in Piccadilly Circus and then in Paris. On her death her genitals were dismembered, preserved, and put on display at the Musée de l'Homme in Paris. They remained there until the early 1990s. An 1885 German book, *Woman: An Historical Gynecological and Anthropological Compendium,* which charted female genital morphology by racial type, was translated into English as *Femina Libido Sexualis* and became so popular it was reprinted 11 times, with the 11th edition being in 1924. The evidence was all bogus, and the images and poses were chosen because they fitted the mythology that had developed. This cultural narrative adds another layer of complexity for Black women when it comes to their sense of self and sexual well-being. In an act of reclamation, Cardi B and Megan Thee Stallion turned this stereotype back on itself in 2020 in their "WAP" video. It's ironic how many white men have been outraged by it.

In England toward the end of the First World War, possession of (or certainly knowledge of) the clitoris was used as evidence of being a German sympathizer. In an Old Bailey case in 1918, actress, dancer, suffragette sympathizer, and possible lesbian Maud Allan sued her accuser, Noel Pemberton Billing, a right-wing MP and editor of the newspaper the *Vigilante,* for his libelous accusations that she was part of the "cult of the clitoris" seeking to advance Germany's cause. The cult allegedly enlisted the wives of powerful men to undermine the British war effort and was in possession of a book that named thousands of German sympathizers. Billing suggested that Allan had proved to the court she was a spy purely by knowing the word *clitoris*—which had been used as a sting, he explained, because it "would only be understood by those who it should be understood by." Billing testified that he had been told by a doctor that the term referred to "a superficial organ that, when unduly excited or overdeveloped, possessed the most dreadful influence on any woman, that she would do the most extraordinary things."[26] It became laughably evident that many of those in court were not familiar with the word. Allan was asked in cross-questioning if she was a medical man, because "nobody but a medical man or people interested in that kind of thing, would understand the term." It is alleged that during the trial Lord Albemarle asked a fellow member of the Turf Club (preserve of the aristocratic male elite, based in Piccadilly, London) who "this Greek chap Clitoris they were all talking about was"—but I'm inclined to think he was being ironic, given that the majority of the club members were "principally engaged in amusing themselves."[27] Maybe he was ignorant, or maybe he was being a wag.

At another point in the trial the prosecution used the word "orgasm," which also caused confusion. The judge requested that the word be repeated, and Allan's chief barrister, Sir Ellis Hume-Williams, asked, "Is that some unnatural vice?"[28] Billing and his cronies were unhinged and driven by motives of revenge and an anti-gay agenda, linking what was seen as unconventional sexuality with moral depravity—but it is notable that the case was deemed worthy of a six-day trial in Britain's foremost criminal court. The jury, reflecting sexual sensibilities of the time, didn't award in Allan's favor. Neither she nor Clitoris had been slandered.

It's a button for alleviating hysteria.
Now the story gets confusing. If you were a woman and happened to live through a witch hunt, your clitoris could indict you as a witch; yet on the other hand it might also have been used to treat "wandering womb," which manifested as a variety of afflictions. Described in the early Greek Hippocratic medical works and by prominent Ancient Roman physicians including Galen, the idea that the uterus could cause any

number of ailments including breathlessness, fainting, lethargy, and convulsions remained popular for centuries. The symptoms were many and shifted, like the problematic womb, depending upon the preoccupations of the day. Anxiety, sleeplessness, irritability, heaviness in the abdomen, headaches, erotic fantasy, and sexual forwardness were all added to the list.

We are often told that the word *hysteria* comes from the Latin *hystericus* or the Greek *hystera,* which it does—but as a catchall for a myriad of symptoms, it is New Latin[29] and was adopted with glee by the Victorians. They knew that the womb didn't literally wander, but were convinced that biology was destiny when it came to women and their uteruses and were thrilled to have it reinforced with new translations of ancient texts.[30]

Hysteria remained a diagnosable condition until the term was dropped by the American Psychiatric Association in 1952. Luckily this gynecologically determined disease came with its own gynecological solution. In her book *The Technology of Orgasm,*[31] Rachel Maines says it was widely believed that these symptoms could be treated by medical stimulation of the clitoris to achieve a so-called hysterical paroxysm. She states that in numerous medical sources from Galen through the 19th century, "massage to hysterical paroxysm" is given as the cure for hysteria. How widespread this practice was is up for debate; but in a chapter on women's diseases in his 1599 medical compendium, *Observationes,* Pieter van Foreest recounts the case history of a 44-year-old widow cured in such a way. The patient was afflicted by *praefocatio matricis,* which translates as "suffocation of the mother," meaning it was thought the widow had retained seed in her womb—another echo here of the idea that women were inside-out versions of men and their bodily functions mimicked those of men, who expelled their seed altogether more efficiently. Foreest's account reports that:

> Because of the urgency of the situation, we asked a
> midwife to come and apply the following ointment
> to the patient's genitals, rubbing them inside with
> her finger . . . For such titillation with the finger
> is commended by all physicians, including Galen
> and Avicenna, particularly for widows and persons
> abstaining like nuns.[32]

The fact that this treatment was also seen as suitable for nuns,
who make a vow of chastity and follow religious rules that
deem masturbation abhorrent, is an indication of how divorced
from the arena of sexual pleasure this kind of stimulation
was. Maines, whose book considers the history of orgasm,
argues that "[s]ince no penetration [with a penis] was involved,
believers in the hypothesis that only penetration was sexually
gratifying to women could argue that nothing sexual could
be occurring when their patients experienced the hysterical
paroxysm during treatment."[33]

Looking up a definition of *hysterical paroxysm,* just to check
we really were talking about an orgasm, I found it defined as
"An obsolete 19th-century term for a female orgasm, when such
was regarded as abnormal."[34] This highlights the ignorance
around female sexual capacity and pleasure as late as the 19th
century. It also tells us something about the sex lives of the
scientists and doctors who labeled female orgasm as "abnormal." I wonder what their wives, if they had them, would have
said on this matter.

Can this treatment really have been a big thing? Isn't this
an example of one piece of factual evidence being blown out of
all proportion, like your nine-year-old telling you that *everyone*
in the class goes to bed at midnight? Maines claims to have
found descriptions of this treatment for female hysteria in the
Hippocratic works from nearly every century, and consistently
from the 15th to the 19th centuries, from multiple sources.[35]

If she is right, then on this occasion it seems your nine-year-old is not exaggerating. Other academics argue that while the references might refer to massage and oil and female parts, we can't tie them down specifically to medical masturbation, or necessarily as having been performed by a third party. We will probably never know.

@its.personalgirls

Many references *are* ambiguous, but while researching this section I came across a widely published and marketed 1840s American book in its 49th edition, by a Dr. Frederick Hollick, called *The Diseases of Women: Their Causes and Cure Familiarly Explained with Practical Hints for Their Prevention for Female Health*. In it the author lists treatments for hysteria, including laudanum enemas and vaginal injections.[36] He strongly advises against "other practices" on account of the "probable moral consequences." He reassures readers that treatment *can* be effected by "less objectionable means." The real hint that he is talking about the treating of hysteria through hysterical paroxysm comes when he cautions against manipulative hysterics and hypochondriacs who are inclined to go "running" to "notorious charlatans" who "excite the mind through the body—the St. John Longs of rubbing celebrity, and the Campbells of celestial-bed notoriety." Hollick argues that these treatments promise "gratification as well as excitement," and for this reason should be avoided.

There is no evidence to suggest that doctors and midwives treating hysteria in this way enjoyed inducing hysterical paroxysms. It was time-consuming, and the business of learning to bring a woman to climax through clitoral stimulation was hard. And—let's be honest—many women take time to discover how to do it themselves before they become mistresses of their pleasure, and their partners struggle too. It's not easy at first. The sweet spot can be elusive and, if one's head is not in the right space, sometimes she can't be summoned at all. "Pelvic massage in gynecology has its brilliant advocates and they report wonderful results," wrote Samuel Howard Monell in 1903, "but when practitioners must supply the skilled technic with their own fingers the method has no value to the majority." He added that "special applicators [he is referring to motor-driven ones here] give practical value and office convenience to what is otherwise impractical,"[37] further advocating that his versatile devices work for both "extra—and intra—pelvic massage." The doctors were not necessarily kinky. It was a job, like sex workers being paid to provide hand relief—they don't report being turned on by the task, either. However, women certainly weren't permitted to have the appliances at home, where they might take control of their own treatment. No, absolutely not. Far too risky.

According to James Manby Gully, M.D., writing about hysteria in *The Water-Cure in Chronic Diseases: An Exposition,* "The douche is a very necessary part of the treatment; and played well on the loins, tends powerfully to facilitate the uterine functions."[38]

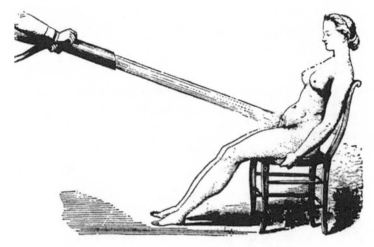

Are we having fun yet? Pelvic douche, c. 1860.

Clearly there is some ambiguity here, because the water is angled toward the patient's lower abdomen and the top of her mons pubis—but if you tipped your chair back slightly and parted your legs, then you'd quite possibly, if you weren't terrified or overcome with embarrassment, have a hysterical paroxysm after a while. I, along with many others, learned about the benefits of a douche played well on the loins from reading a tatty old copy of Erica Jong's 1973 *Fear of Flying* that I picked up in a travelers' hostel in the early 1990s, and it's something all showerhead designers should get their heads around.

Maines establishes the longevity of hysterical paroxysm as a treatment for hysteria, and while I don't know how widespread it truly was, I'm fascinated by the notion that it happened at all.

If only the women of those eras could bear witness—though as I write this I am reminded of how problematic it continues to be for women to bear witness to their sexual experiences. If in any doubt, read Lucy Crawford's memoir, *Notes on a Silencing*, or Jodi Kantor and Megan Twohey's *She Said*.

I wanted to know how widespread hysteria was as an illness. In the 17th century Thomas Sydenham, an English physician posthumously referred to as the English Hippocrates and renowned also for documenting gout and cholera, stated of hysteria that "women, except for those who lead a hardy and robust life, are rarely quite free from it." It was, he said, the disease that occurred most frequently, estimating that it accounted for "one sixth of all human maladies,"[39] and it was still going strong in the late 19th century. Russell Thacher Trall, who published copiously about family health and "hygienic" practices, estimated that in America "more than three-fourths of all the practice of the [medical] profession are devoted to the treatment of diseases peculiar to women," asking, "How can the doctors afford to have the women healthy?"[40] For those with an eye to business, this group clearly constituted a wonderful marketing opportunity. One can see how the newly branded "therapeutic services" market was born. The technology of the Industrial Revolution that had given us steam engines and electricity could be put to work to create water jets and vibrating devices to treat, among other things, hysteria.[41] Maines controversially argues that these were early vibrators. Again, the evidence is unclear. Were they vibrators, as in sex aids, or vibrating therapeutic aids that could also be used as a sex aid, like the vibrating toothbrushes you can buy in packs of two off the shelves in Walgreens? I'm all for normalizing women's masturbation, but any treatment that linked orgasm to illness rather than sexuality—not to mention the thorny question of consent—would be a contentious and dark point in the history of the clitoris.

That is not to say all physicians were blind to the orgasmic potential of therapeutic massage. It seems there had always been some understanding of what was going on. Foreest, referenced earlier, who was specific in his directions about how to induce a genital cure for his widowed patient, also advised of the treatment that it was "less often recommended for very young women, public women, or married women, for whom it is a better remedy to engage in intercourse with their spouses," and Dr. Hollick concluded his chapter on hysteria with, "An opinion prevails very generally, that in all these cases marriage is advisable, and in the great majority this is perhaps true."[42] In her research, Lana Thompson found a number of 15th and 16th century European physicians and thinkers pondering the ethics of treating women "by titillation and friction of the genital area."[43] Curran Pope, in 1909, wrote in his hydrotherapy manual that hysteria and other female ailments could be caused by "imperfect or unsatisfactory intercourse" and that in these instances hydrotherapy applied "to the inner surfaces of the thighs" was recommended. "Douches are, as a rule, more agreeable to the majority of individuals than the other forms of hydriatic procedure . . . It sets the tissues in a vibration impossible to describe; experienced, it is never forgotten."[44]

However, these doctors were up against an establishment that could not contemplate the possibility that intercourse and penetration might be unsatisfying. Wilhelm Griesinger (1817–68), a German physician, claimed in his book *Mental Pathology and Therapeutics*[45] that hysteria definitely wasn't caused by sexual frustration because of its "great frequency among married women ... and the frequency of the affection among prostitutes." This is akin to arguing that wild animals like living in cages because otherwise why would there be so many in zoos? It's a logical fallacy, arguing from adverse consequences. Science was surprisingly oblivious to the evidence because it was blinded by the meta-idea that sexual intercourse was, by its very nature, fulfilling and highly sought after. Sexually active women not experiencing paroxysms of pleasure from a penis? Impossible! It's the blind men and the elephant all over again.

The blind men and the clitoris

@its.personalgirls

The vagina as orgasmic powerhouse

The vagina-as-inverted-penis model of a woman's sexual anatomy led to wild assumptions about how the vagina worked as an organ of sexual pleasure. If you believe that the vagina is an inverted penis, then you might imagine that with a few strokes it could be aroused. The long-held insistence that insertion of a penis—or indeed anything—into the vagina would trigger desire and orgasm is extraordinary.

The anxious male medical profession was consumed with concern about the introduction of the speculum as an instrument of pelvic examination in the mid-1800s, and a meeting at the Royal Medical and Chirurgical Society of London to discuss the threat the speculum posed to society was so packed that many attendees had to stand. Previously any pelvic examination would have been under the cover of skirts, petticoats, or sheets; there would definitely have been no looking. Robert Carter, a British physician and social critic, wrote about young unmarried women being "reduced by the constant use of the speculum, to the mental and moral condition of prostitutes; seeking to give themselves the same indulgence by the practice of solitary vice; and asking every medical practitioner under whose care they fell, to institute an examination of the sexual organs."[46] There are so many things wrong with this statement. Most women who have had pap smears will find the idea of the speculum being a source of sexual stimulation laughable. I'm outraged by the belief that prostitutes (although notably not their clients) were morally depraved and sex-crazed, and I'm dismayed but not surprised that orgasm was considered a "solitary vice." More to come on this later, but when did masturbation become less taboo for men than women? Do we really believe that among the men who were so against it in women there weren't some who indulged in the vice themselves? Consistently, research studies on the prevalence of masturbation find that somewhere in the region of 90 percent of men claim to masturbate. Were Victorian men so very different?

Even *The Lancet,* one of the world's oldest and most respected peer-reviewed medical journals, apparently ran a piece in 1881 that claimed gynecological examinations could "ignite sexual passions in women" and encourage them to "satisfy their own lusts,"[47] although while I read this in a number of secondary sources, I couldn't locate the original article myself. Belief in the power of the penis or anything penis-shaped to satisfy women, just through insertion into a vagina, also plays out in the history of the design of vibrators—once, of course, they stopped being vibrating therapeutic aids. Consider also the heterosexual sex most frequently shown in TV shows and movies: missionary-position coitus beneath the covers, hands-free, with happily orgasmic women.

Therein lies another book entirely, already written by Elaine Showalter[48]—but given the misunderstanding and repression of women, is it any wonder they suffered from hysteria? We are only just beginning to understand the impact on mental health for people who experience physical vulnerability, poverty, lack of education, misdiagnosis, or loss of control over their own lives, not to mention sexual assault. Go figure.

We will leave our story here for a moment. The next section backtracks to explore a parallel history where the clitoris *was* acknowledged as the seat of sexual delight.

Or . . . the Clitoris Is a Penis.
Can women be trusted with it?

For trans men taking testosterone, this is a self-evident statement. The clitoris can become a neo-penis, as tissue enlargement often occurs with testosterone hormone therapy. However, this book is concerned with unraveling the history of our dear little clit, and its marginalization by those who didn't believe in its power. Through history, some voices consistently positioned the clitoris—not the vagina—as the woman's equivalent to a penis. Hippocrates (460–375 BCE) used the term *columella* or "little pillar" to indicate the clitoris, implying that he understood its erectile nature. What a shame that his observations weren't taken as seriously as his oath, and that his diagnosis of "hysteria" was the medical discovery that stuck. Similarly, Avicenna (980–1037 CE), another founding father of early modern medicine, named the clitoris the *albatra* or *virga,* meaning "rod." Albucasis, a great medieval Arab physician and father of surgery (936–1013 CE), in his encyclopedia of medicine, named the clitoris *tentigo* (tension), which suggests he understood her orgasmic properties. His encyclopedia was translated into Latin, became standard reading throughout the medieval period, and was still being reprinted in the 1770s. The medieval books read like jabberwocky to me so I am not going to list all the references here, but Henry Daniel, whose surviving texts outnumber those of Geoffrey Chaucer's *Troilus and Criseyde* by two to one, attesting to his writings' significance, is one example of a writer who acknowledged the clitoris's erectile and sexual potential. The medieval and early modern medical and philosophical communities *did* have access to an array of anatomy texts that included the clitoris, and among those who viewed the clitoris as a counterpart to the penis there has always been a belief in the capacity of a woman for sexual pleasure as strong as a man's.

Later examples of clitorati are Helkiah Crooke, court physician to King James I, and French physician Nicolas Venette, who addressed couples in the form of a marital advice manual in 1687, advising them that the clitoris was the "Fervor and rage of Love...It is there that Nature has placed the throne of her pleasures and voluptuousness, as she has done in the male gland."[49] There were other proponents, such as the French scholar Jean Riolan and the Danish anatomist Thomas Bartholin—famous for discovering the lymphatic system—but unfortunately the clitoris lacked many advocates. Given the status of women at the time, this is not surprising; women weren't able to speak for themselves, and professional men had other avenues to research that weren't as controversial. They always had an eye to their sponsors.

The disappearing clit

The clitoris was often found and then hidden or lost. Charles Estienne (1504–1564) "discovered" it, but he related its function to urination.[50] That's still a common misunderstanding, but it seems Estienne might have played up this confusion, as he also said the clitoris was a "membre honteux" (shameful member) and should not be shown in texts. He must have known what the clitoris was about and been squeamish about female sexual pleasure, otherwise he wouldn't have deemed it shameful. But not including it on diagrams? That's censorship.

When Realdo Colombo (1516–1559) tried to put the clitoris—albeit by another name—on the anatomical map, his science was scoffed at. Colombo studied and subsequently worked under Vesalius at the University of Padua, and ultimately succeeded Vesalius as master of anatomy and surgery in 1544. Possibly aided by Mrs. Colombo, if there was one, he made discoveries that contradicted Vesalius, including arguing for the existence of the clitoris, which he described as "the principal seat of women's enjoyment in intercourse." He further observed that it functioned similarly to a penis, writing that "if you touch it, you will find it rendered a little harder." Colombo was the first Renaissance anatomist to recognize the clitoris's role in sexual pleasure for women and to put it out there. Having described the clitoris's location, he writes:

> And this dearest reader is that, it is the principal seat of women's enjoyment in intercourse; so that if you not only rub it with your penis but even touch it with your little finger the pleasure causes their seed to flow forth in all directions, swifter than the wind, even if they don't want it to ... since no one else has discerned these processes and their working; if it is permissible to give a name to things discovered by me, it should be

called the love or sweetness of Venus. It cannot be said how much I am astonished by so many remarkable anatomists, that they have not even detected it on account of so great advantage this so beautiful thing formed by so great art.[51]

Once I get over the claim that Colombo thought no one had ever discerned the clitoris—I mean, did he really think not one woman had discovered it for herself?—and I remind myself to set Colombo's writing in context, which was for a scientific community made up exclusively of 16th-century men, then I love him for the enthusiasm and unabashed pleasure he took in his groundbreaking discovery and I wish his name, *the love or sweetness of Venus,* had stuck.

In an era that historians call the golden age of discovery, as mankind's understanding of the world flourished, laying claim to the discovery of what we now call the clitoris became *a thing.* First, Colombo announced with great authority that he had identified it. He was challenged by fellow anatomist, Gabriele Falloppio, who claimed in 1561 that *he* had first found this "hidden" and "neglected" part of a woman's anatomy, saying that "if others have spoken of it, know that they have taken it from me or my students."[52] And there ensued an undignified row about who discovered the clitoris first.

Either way, it didn't matter much, because Vesalius's authoritative response to their dispute was thoroughly dismissive.

It is unreasonable to blame others for incompetence on the basis of some sport of nature you have observed in some women and you can hardly ascribe this new and useless part, as if it were an organ, to healthy women. I think that such a structure appears in hermaphrodites who otherwise have well-formed genitals, as Paul of Aegina[53] describes, but I have

never once seen in any woman a penis . . . or even the
rudiments of a tiny phallus.[54]

For the influential and esteemed Vesalius, a clitoris was a quirk
of nature, "some sport," with *sport* being used in the old-fashioned way to mean a game or joke. He believed it was "useless"
and definitely not associated with healthy women.

It seems incredible that Aristotle, Galen, and Vesalius missed
the clitoris—but it's more understandable when you realize
they had not actually looked at the bodies of women, dead or
alive, except perhaps those, if any, that they were intimate with.
They examined more readily available animal carcasses and
extrapolated from there. And if cadavers were available, they
may have been too prudish to look for the clitoris. Perhaps,
too, the women in their lives were shy or ignorant themselves.
As I've mentioned, it's an unspoken truth that many women
take a long time to discover that their clitoris plays a vital role
in sexual pleasure. If these scientists were alive today, they
might argue that the clitoris was hidden and hard to get at.
Maybe even that their cadavers were all elderly and suffering
from shrinkage. I believe they were so fixated on the belief that
woman was an inside-out man that they looked no further.
Society was not much interested in or open to female sexuality
anyway, so despite the best efforts of Estienne, Colombo, and
Falloppio, the clitoris's discovery for the world of science was
put on hold for another century and another anatomist.

Just over one hundred years later, in 1672, it was the turn
of Regnier de Graaf to discover the clitoris, and he gave her
the name that would stick. He signaled the need to distinguish the labia from the clitoris and, he said, to avoid future
confusion, "always give it the name, *clitoris*." Latin or Ancient
Greek were traditionally used in scholarly scientific writing as

labels, because even in the 16th and 17th centuries academics wrote in Latin. If there wasn't a ready-made word to hand, or a word had been lost, it was made up, like the names for magical creatures and spells in Harry Potter. This is still the case: in 2011 a group of Australian scientists named a new and rare species of horsefly *Scaptia beyonceae* on account of its gorgeous booty. It was only with 17th-century advances in printing and the subsequent wider circulation of texts that these labels became fixed reference points, which is probably why we lost earlier terms like *tentigo*. Since De Graaf, *clitoris* has remained the scientific label.

"We are extremely surprised," wrote De Graaf (using the royal *we*), "that some anatomists make no more mention of this part than if it did not exist at all in the universe of nature. In every cadaver we have so far dissected, we have found it quite perceptible to sight and touch."[55] Bless him, De Graaf gave a comprehensive account of clitoral anatomy, including some of the internal aspects, notably the bulbs. I'm puzzled by one thing, though: how come the clitoris-related work of Estienne, Colombo, and Falloppio was forgotten or lost when their work on the urethra, blood vessels, and Fallopian tubes survived? Odd, that.

I am also entertained by De Graaf's comment that "[i]f these parts of the pudendum had not been endowed with such an exquisite sensitivity to pleasure and passion, no woman would be willing to take upon herself the irksome nine-months long business of gestation, the painful and often fatal process of expelling the fetus and the worrisome and care-ridden task of raising children."[56]

A female perspective

One of the most interesting accounts of the clitoris in English, although that name wasn't in use yet, comes from a 1671 midwifery book by Jane Sharp. Significantly, this is a text written by a woman for women, so it gives us insight into an otherwise unrecorded female world and raises the possibility of an unwritten subculture that might have existed around female sexuality. Male physicians and scientists were recording their beliefs about everything for posterity, but we know very little about the conversations that were taking place in female circles.

Before doctors (men by default until Elizabeth Blackwell in 1849) felt the need to take responsibility for childbirth, it was firmly situated in the domain of women, and Jane Sharp provides a glimpse of what might have been circulating within these female-only spaces. Her book, *The Midwives Book, or the Whole Art of Midwifry Discovered*,[57] addresses its readers as "sisters." Sharp's tone is intimate, caring, and one of experience in the birthing room. In her essay on the impact of this book, Catherine Morphis argues that it was almost certainly intended for a wider audience.[58] Lisa Cody, another academic expert on Sharp, argues that "[g]iven its hundreds of pages on genital anatomy, conception, marriage, female diseases, folklore and various matters not immediately connected to the delivery of infants, the book seems designed as much for curious mothers and fathers as for midwives."[59] Sharp's audience may have been wide and diverse and the title of her book a clever ploy to foil the censorious, as the mention of midwifery would have deterred serious-minded men of science or religion from engaging with it. Women's work.

Sharp described the clitoris as "but a small sprout, lying close hid under the Wings [labia], and not felt, yet sometimes it grows . . . and it will swell and stand stiff if it be provoked." She described it as being sinewy, spongy, and hard of body. She commented that it could rise and fall like a man's penis (or "yard") and advised that it "makes women lustful and delight in Copulation, and were it not for this they would have desire nor delight."[60] Finally, and monumentally, Sharp acknowledges that it is a combination of the stirring of the clitoris *and the imagination* (my italics) that arouses a woman most successfully. This is a remarkable insight and an area of understanding about sexual arousal, desire, and response not returned to by the world of science or psychology until relatively recently. Her emphasis on the clitoris is perhaps not surprising as it was widely believed, as detailed below, that female orgasm was essential to conception, so in a book dedicated to childbirth, ensuring this first step would be critical.

Jane Sharp's work is an example of what academic Helen Lefkowitz Horowitz calls "vernacular sexual culture."[61] Horowitz reminds us that when it comes to sex, people can hold multiple perspectives, and what is regarded in academic and decision-making circles as the dominant belief or code of behavior may not in reality be what is happening on a mass level. The incidence of masturbation would be a case in point.

In 1684, another book appeared in English (rather than the academic language of Latin) that ran to hundreds of editions, right into the late 19th century: *Aristotle's Masterpiece.* Maddy Smith, a curator at the British Museum in London, which holds a copy, says it's "a book for the common people that would've been cheaply printed, sold 'under the table,' and hidden under the mattress at home. With its advice for both men and women, it would've been furtively rifled through." The book offers advice on everything from marriage and sex to the "greensickness" in virgins and how to get a male or female child. It also covers monstrous births, midwifery, and the wandering womb. It is a mash-up of early-17th-century medical works and popular old wives' tales, authored by a pseudo-Aristotle who saw a gap in the market. The anonymous author, thought to be male because of the inclusion of a poem about intercourse written from the perspective of a penis, advocates strongly for sexual gratification for both partners in intercourse, implying that it is natural to experience sexual pleasure: "For Nature does in this great work design/Profit and Pleasure, in one Act to join."

Of orgasm, the text says: "Much delight accompanies the ejection of seed, by breaking forth of the swelling spirit, and the stiffness of Nerves," adding that the experience for women is the same. The metaphor of the seed applied to both sexes, as it was thought that both men and women emitted seed at the point of orgasm. Indeed, until the reproductive cycle was fully understood in the early 1800s, it had been believed by most academics and physicians since classical times that it was the event of orgasm that triggered conception. While women were credited with only providing the substance for the growing of a baby (which is why they didn't bleed during pregnancy or lactation), men were believed to provide the impetus. It is worth noting that despite academics squabbling over the clitoris, the populace at large believed in the importance of female orgasm and their go-to sources told them how to experience it.

Chapter XIII of *Aristotle's Masterpiece* is dedicated to "The External and Internal Organs of Generation in Women," and here the author is very clear that knowledge about the secrets of a woman's anatomy is important.[62] "It is absolutely necessary that they should be known for the public good." On the clitoris, it reads:

> The *Clitoris* is a substance in the upper part of the division where the two wings meet, and the seat of venereal pleasure, being like a man's *penis* in situation, substance, composition and power of erection, growing sometimes to the length of two inches out of the body, but that never happens except through extreme lustfulness or some extraordinary accident. This *clitoris* consists of two spongy and skinny bodies, containing a distinct original from the *os pubis*, its

tip being covered with a tender skin, having a hole or passage like a man's yard or *penis*, although not quite through, in which alone, and in its size it differs from it.

Chapter XV, "A Description of the Use and Action of the Several Generative Parts in Women" says: "The action of the clitoris in women is similar to that of the penis in men, viz., *erection*; and its lower end is the glans of the penis, and has the same name. And as the *glans* of man are the seat of the greatest pleasure in copulation, so is this in the woman."

Strikingly absent in the descriptions of female genital anatomy, which also include paragraphs on the labia, cervix, and womb, is any mention of the vagina as a specific site of sexual pleasure beyond the manner in which it is dilated with the "pleasure of procreation." The vagina is presented as the sheath for the penis, a means of giving the thrusting penis protection on its passage to the womb, but it is the clitoris that is lauded. Finally, couples are advised to excite their desires mutually before they begin conjugal intercourse. This 17th-century under-the-table narrative about sex has a strong emphasis on mutual pleasure that is noticeably lacking in many of today's sex-ed books and in too many sex-ed classes.

One can imagine why this book was so popular, with practical advice on conception and birth, as well as the titillation and encouragement provided by the sections on anatomy. The advent of printing and the publication of the Bible in English in the previous century meant that by the middle of the 1600s nearly half of the population could read. *Aristotle's Masterpiece* quickly found its way across the Atlantic; within a year, by 1685, there were eight copies in the inventory of Boston bookseller Chiswell, and it became a popular text. In 1744 a certain minister, a Jonathan Edwards from Northampton, Massachusetts, led a formal inquiry among his flock about their reading of it.

He was worried that the *Masterpiece* would defile congregants and make them unsuitable for receiving communion. Jonathan Edwards's own manuscripts[63] give us an insight into what happened, and it is a scenario that could have played out just yesterday. Some girls reported that a group of boys had been bragging and taunting them about what they knew, saying things like, "We know as much about ye as you know, and more too." There ensued a parish manhunt, and eventually a small gang of boys confessed to sitting up late reading the book. Unrepentant, one declared, "Don't you think I would sit up again if I had the opportunity?" Of course he would!—in the same way that 90 percent of today's American college men viewed porn in the past year[64] and many younger teens cite wanting to learn about sex as their reason for watching it.

The clitoris is finally put on the anatomical map.
Despite the popularity of books like *Aristotle's Masterpiece,* male academics continued to play their centuries-long game of hide and seek with the clitoris. In 1840 another scientist, Georg Ludwig Kobelt, claimed to have discovered it. Perhaps this should have been the topic of my timeline, a timeline of discoveries and losses; a history of careless academic mislaying and casual forgetting, like a schoolchild with their gym kit or lunch box. "I have made it my principal concern," said Kobelt in an essay, "to show that the female possesses a structure that in all its separate parts is entirely analogous to the male," but he goes on to lament that "our knowledge in regard to these female structures is still full of gaps."[65] Kobelt's account and diagrams of the clitoris are detailed and he mapped it thoroughly, showing that the clitoris was comprised of much more than the visible glans and hood which make up part of the external vulva.

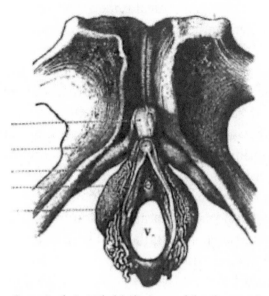

Georg Ludwig Kobelt's diagram of the clitoris, 1844.

Kobelt performed dissection, comparative anatomy, and injection studies: real dissections on real women, not just on monkeys and pigs. Kobelt identified the clitoral glans, shaft, bulbs, and associated muscles, and he understood that the whole was a cluster of erectile parts extending into a woman's pelvic cavity. He identified the blood supply and wondered at the thickness of the nerves as they enter the glans or visible part of the clitoris. "Here they are, even before their entrance so very thick that one scarcely imagines how such an abundance of nerve mass can still find room between the countless blood vessels of this very tiny structure." He even goes so far as to say, "We can grant the vagina no part in relation of the specific pleasurable sex feelings in the female body." There you have it, a practically full anatomical understanding of the clitoris, in 1840. Finally!

The brilliant Helen O'Connell and her fellow researchers —to whom I will return in a chapter of their own because their research takes anatomical understanding of the clitoris one step further—lamented in their 2005 essay that Kobelt's descriptions "should have guided the authors of anatomical textbooks to provide accurate information, but that has not been the case. The typical anatomical textbook description lacks detail, describes male anatomy fully and only gives the differences between male and female anatomy rather than a full description of female anatomy."[66] Twentieth and even most 21st-century anatomical text books continue to focus on woman's reproductive anatomy and downplay the clitoris, with any discussion of clitoral function being rare, despite the scientific knowledge available. There is also a trend in anatomical and sex-information books to position sexual anatomy images consecutively, so that parallels can be drawn between the successful and size-dominant penis and the tiny clitoris, reinforcing the age-old fallacy that woman is a lesser version of man.

Fear of masturbation, the solitary vice
The knowledge that the clitoris was the seat of sexual pleasure brought with it a dread of what women might do with it when they were not attended by a man. Women, remember, were not supposed to be independent thinkers or doers, and as such were definitely not to be trusted with their sexuality—just as they could not be trusted with the vote. Now it was as if the world had worked out that women were not poor versions of men, but were an entirely separate, possibly opposite, entity. As we've seen, a whole host of thinking pursued this line of thought in other ways, using man as the benchmark and woman as the opposite. Men were active, women passive. Men were rational, women irrational. Men were strong, women weak. Now they were not lesser men or parallel to men, they were different! And what is different is to be feared. All this at a time when

the Industrial Revolution was changing the social landscape for women as well as men, opening up an industrial world of city living, factory work, and shop work for an ever growing working and middle class, with the increasing freedom to travel facilitated by railroad expansion. This was going to be a nightmare for the patriarchy!

In addition, the Western world was expanding through colonialization, and fear of the "other" loomed large. Western man, who had been on his own incredibly elite journey of enlightenment, feared slippage or a primeval regression. Looking to the colonies, these men were able to cite the people they termed "savages" who were living there as examples of a more primitive, more sexualized, less evolved, people. The women in those lands were framed as examples of the latent lusty potential of women, if left to revert. Even though it was not Darwin's intent, his theory of evolution (published in 1859) was quickly hijacked to chart white man's genetic progression to greatness. His cousin Francis Galton championed eugenics, including female sterilization, as a means of breeding out "undesirable" traits, as the horror of white males losing dominance became a concrete anxiety. Darwinian evidence could be summoned to justify man's responsibility for preserving his enlightenment progress. "The chief distinction in the intellectual powers of the two sexes," Darwin wrote, "is by man attaining higher eminence, in whatever he takes up, than women can attain—whether requiring deep thought, reason or imagination, or merely the use of the senses and hands."[67]

Oh, what women might do with their hands. Masturbation had always been a religious no-no, but the 18th and 19th centuries saw it become a medical concern. In 1706, Samuel Auguste Tissot, a Swiss doctor, had claimed with great authority in his book *L'Onanisme* that masturbation would spiral you into a physical as well as a moral decline, but 1756 saw masturbation angst kick off big-time with the publication of *Onania: or, the heinous sin of self-pollution and all its frightful consequences (in both sexes) considered*. This widely distributed pamphlet told readers that masturbation made you blind, epileptic, hysterical, consumptive, and/or barren. In an era when science was discovering diseases faster than it could develop cures for them, masturbation was a useful culprit and one already singled out by religion as evil. Further ailments were attributed to the solitary vice: vomiting, paralysis, back pain, pimples, memory loss, attacks of rage, and fever.

As an activity, it was going to take a long time for masturbation to receive a clean bill of health—despite assurances in 1897 by the widely published physician, progressive thinker, and social reformer Henry Havelock Ellis that there was no evidence linking masturbation to any serious mental or physical disorder.[68] I would add, however, that Ellis claimed masturbation was, after adolescence, "[a]pparently more frequent in women."[69] It only ceased to be a diagnosable condition in America in 1968 when it was removed from the *Diagnostic and Statistical Manual of Mental Disorders,* published by the American Medical Association, who finally declared it to be "normal" in 1972 in their publication *Human Sexuality.* Still, it remained controversial. In 1994 U.S. Surgeon General Dr. Jocelyn Elders suggested it should be mentioned as safe and healthy in school curricula. In answer to a question at the UN Conference on AIDS as to whether she thought teaching about masturbation in schools might reduce unsafe sex, she replied: "I think that is something that is part of human sexuality and it's part of something that should perhaps be taught. But we've not even taught our children the very basics." She was forced to resign by Bill Clinton as a consequence. His chief of staff, Leon Panetta, said, "There have been too many areas where the President does not agree with her views. This is just one too many."[70] Sex is clearly political as well as personal.

In the U.K. in 2007, three planned documentaries about masturbation, set to be screened during a so-called Wank Week, were pulled because of concerns about a state-owned, advertising-funded TV channel being seen to promote such activity. The channel's chief executive criticized the project as pandering to the "obsession with adolescent transgression and sex." Masturbation, a transgression? Surely it is one of the oldest and most widely pursued activities known to man? (I'm using *man* in the collective sense here.)

So, despite understanding the clitoris's anatomy and capacity for sexual pleasure, 18th- and 19th- century society was terrified of it. Unlike the penis, which provided semen—essential for procreation—the clitoris apparently served no function. Masturbation was threatening because it was a solitary and secret recreation, it relied on the imagination (who knew what was going on in women's dangerous little heads?), and the desire was potentially endless (who knew when they'd stop?). By the end of the 19th century it was commonly believed that a large clitoris was both a sign of the vice and the result of it, with *clitorism* entering *The Medical Lexicon: A Dictionary of Medical Science* in 1854, defined as "an unusually large clitoris," with the qualifying information "the abuse made of the clitoris to satisfy an unnatural sexual desire" being added to the 1900s edition.[71] There was also a risk that its engorgement was not the result of solitary activity. Suppose women had been (shock, horror) doing it with each other?! Now that the vagina was being sidelined as *the* site of sexual pleasure, maybe the penis would get sidelined too?

What's more, women were being educated and beginning

to have the means of earning their own money, and as many as ever were dying as a result of childbirth. (Despite advances in rural midwife care, the shift to hospital deliveries with increased intervention in uncomplicated deliveries kept maternal death rates at a steady high until the 1930s on both sides of the Atlantic.[72]) And if women could find sweet joy unaided, without a penis, what was the point of men?

One of many fears that gained ground in the 19th century was that if women experienced orgasm through masturbation they would, as a result, suffer from "marital aversion."[73] Taking responsibility for protecting women from external titillation became something of a Victorian and Edwardian obsession, and in an era famous for industrialization and expanded forms of transport, the risks were numerous. There was great alarm about situations like working a two-foot treadle sewing machine[74] or riding a bicycle, and what they might do for a woman.[75]

Ever wondered why the traditional woman's bicycle had upright handlebars? It was to tilt the sitting position back so the clitoris wouldn't come into contact with the seat. I thought the design was all about comfort, but in the early days of bicycle production, saddles were built to avoid arousal, not discomfort, and as such were branded "hygienic." *Hygienic* seems to have become a byword for not being sexually stimulated, leading to, I suppose, the opposite concept that if one was stimulated it was in some way unhygienic and thus dirty.

In 1892 Edward Ely Van de Warker wrote an article for the *Georgia Medical Companion* on the "Effects of Railroad Travel upon the Health of Women." Such was the fear of women experiencing a thrill down there, without a man in the mix. I now understand categorically that galloping on horses and traveling on trains in 19th-century fiction is about much more than getting from A to B. Implicit within these means of transport is the implication that the women indulging in them are experiencing heightened sexuality. Knowing this, we can reread *Madame Bovary* and *Anna Karenina* with a fresh perspective. All those 19th-century novels where trains are a trope deserve reevaluation.

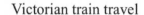

Victorian train travel

@its.personalgirls

I have just turned to the footnotes about the two-foot treadle sewing machine in Maines's book. I share the list of references below both as an example of the obsession with the danger of turning women on and because the image of all these men—so worked up about the possibility of women achieving stimulation through the pedal motion of an old-fashioned sewing machine—is beyond funny. For goodness' sake, you might as well say that walking is dangerous for women on account of the potential for rubbing.

"The Hygiene of the Sewing Machine." A.K. Gardner, *American Medical Times* 1 (1860).

"Influence of Sewing Machine on Female Health." *New Orleans Medical and Surgical Journal* 20 (November 1867).

"On the Influence of the Sewing Machine on Female Health." J. Langdon and H. Down, *British Medical Journal* (1867).

"The Sewing Machine Problem as Seen Through the Pages of the *American Journal of Obstetrics and Diseases of Women and Children.* Charles H. Hendricks, 1868–1873." *Obstetrics and Gynaecology* 26 (1965).

Female Hygiene: A Lecture Delivered at Sacramento and San Francisco by Request of the State Board of Health of California. Horatio Robinson Storer (1872).[76]

According to one authority it was well-known in French workrooms that you could spot an orgasmic seamstress by the sound of a machine being worked for "a few seconds of uncontrollable rapidity." Apparently this sound was to be frequently heard, although the happy sighs were drowned out.[77] It was all that supervisors could do to keep it under control. Can you imagine if Health and Safety had been a thing back then? Think of the risk assessments that would have been required! Didn't these men have better things to do with their time than imagine multitudes of happily orgasmic women in every facet of their lives? The idea did their heads in, and many were keen to police the hell out of it.

One man who took this to the extreme was Dr. John Harvey Kellogg of early Cornflakes fame, before his brother split from him and founded the company that exists today. As a passionate advocate for "pure" living and an ardent anti-masturbator, Kellogg recommended that parents always be on the alert for signs of masturbation in their children, including creeping up on them just after bedtime, ripping back their covers to catch them in the act, and examining their genitals for any arousal. His cure for girls? Applying carbolic acid to the clitoris as "an excellent means of allaying the abnormal excitement."[78] Kellogg's *Ladies' Guide in Health and Disease* recommended the following for treating both nymphomania and "the disposition to practice self-abuse":

> Cool sitz baths; the cool enema; a spare diet; the application of blisters and other irritants to the sensitive parts of the sexual organs, the removal of the clitoris and nymphae, constitute the most proper treatment . . . In an extreme case of this kind brought to us for treatment a few years ago, we were compelled to adopt the last-mentioned method of treatment before the patient could be cured.[79]

Such was the level of misunderstanding and anxiety about the clitoris and women's orgasms that it's no wonder some women took to the "passionlessness" movement, which flourished in 1900s America. I'd love to see these men go up against the formidable Betty Dodson, whose outspoken and infectious advocacy of sexual pleasure and masturbation has been formative for many women since the 1970s. In a heavenly faceoff, I'd back Betty.

It's an obscenity! The rise of censorship and erasure of the clitoris in literature

During the 1800s, a time when there was increasing knowledge about the anatomy of women and reproduction, both the U.S. and the U.K. saw a strengthening of their obscenity laws to restrict the kind of sexual information that was widely available. There had long existed a conviction that licentious erotic literature should not be allowed, on account of its corrupting

potential. In 1727 Britain's Star Chamber prosecuted Edmund
Curl for publishing a book that went by the name *Venus in
the Cloister; or the Nun in Her Smock.* (No prizes for guessing
the plot.) The attorney general argued that the book could
"corrupt the morals of the Subjects of this Kingdom" and that
the "wicked" Curl should be punished for his "Offence against
Morality."[80] Francis Ludlow Holt, whose book *The Law of Libel*
was published on both sides of the Atlantic, argued furthermore that "obscene writings, speakings, and exhibitions" had
a tendency to "disturb the peace and economy of the realm."
Not only were morals at stake, but the black-market economy
was thriving—and both were a civic concern. This might not
have been such an issue for the clitoris given the plot tropes of
the genre, but morally minded men began extending these laws
to include nonfiction books written for the popular market.

Charles Knowlton's *Fruits of Philosophy, or the Private
Companion of Young Married People,* which he circulated
discreetly in his community in Ashfield, Massachusetts, in
1831, argued that sex was a natural instinct and a source of
happiness. One reviewer was appalled and wrote indignantly
that this kind of thinking would "turn the world into a universal brothel."[81] Knowlton was prosecuted and fined. In 1833,
a second version was published by one Abner Kneeland in
Boston, presumably on account of unfulfilled demand for the
text. Knowlton was arrested for a second time and imprisoned
with hard labor for three months. Kneeland subsequently faced
a blasphemy trial for publishing the book along with other
free-thinking literature. It's hard to know whether it was the
information about the female anatomy apropos birth control
or the information about the female sexual response and the
clitoris being acknowledged as "the principal seat of pleasure"
that provided more anxiety. Either way, the establishment

sought to repress the information—unsuccessfully, it seems, as is the way with most bans. The book reappeared in London in 1877, where Charles Bradlaugh and Annie Besant were also tried for publishing it. The publicity from the trial saw sales of the book soar.

I'm curious about Knowlton's use of the phrase that Colombo had applied when he discovered the clitoris in 1599: "the principal seat of pleasure." Where did Knowlton come across it, given that it was originally written in Latin? Who translated it, and in what underground circles had it survived, given that it had been erased from the academic anatomical canon? When did it lapse from our vocabularies? Or was it just coincidence that he used the same words?

In 1870 in Philadelphia, Simon M. Landis, a progressive clergyman and self-styled doctor, was brought before the courts for self-publishing *A Strictly Private Book on Marriage: Secrets of Generation*. His lawyer's defense was that it was not an obscene book because it was neither written to appeal to erotic male interests nor sold in secret (an indirect comparison to the nun book). *Secrets of Generation* was, he argued, written for public enlightenment: it was a medical and scientific work.[82] According to the judge, however:

> Publications of a character which is strictly scientific, strictly medical, containing illustrations exhibiting the human form, if wantonly exposed in the public markets, and publicly advertised for sale in such a manner as to create a wanton and wicked desire for them . . . would be obscene and libelous.

Landis was sentenced to a year's imprisonment, and the campaign by the ruling elite to suppress information about female sexuality and female fertility triumphed again.

Edward Foote's *Medical Common Sense*, first published in

1858 by Wentworth, Hewes & Co., Boston, which equated the "glans-penis of the male and the clitoris of the female," met a similar fate. It originally sold more than a quarter of a million copies and upon being reissued in 1870 as *Plain Home Talk* sold another half a million but in 1876 Foote was given a 10-year suspended sentence under the obscenity laws and forced to remove some content. Anthony Comstock—secretary of the New York Society for the Suppression of Vice, and United States postal inspector—who had persuaded Congress in 1873 to make it illegal to produce, publish, or deliver by mail any "obscene, lewd or lascivious" material, claimed that Foote's *Plain Home Talk* and its advertisement fell afoul of his laws and was "an incentive to crime against young girls and women."[83] Thus we can see how girls and women were "protected" from information about their genitals. Luckily for Amazon (and this book), Comstock's laws prescribing the types of books one can send by mail were essentially repealed in 1971. Given the history of the suppression of sexual information for the masses, and the trend it set for subsequent generations, perhaps it's not surprising the general public became unfamiliar with the clitoris. One (male) half of the population could live in complete ignorance of its existence, and the other (female) half remained ignorant, ashamed, or wisely mute about it.

Things were not much better in the 20th century, particularly in the U.S. In the late 1990s, to be eligible for any of the federal funding set aside for sex education, public schools in America had to teach that marriage was the *only* acceptable sphere for physical relations, and that sex outside of it at any age would lead to irreparable physical and emotional damage. "By 1999 40 percent of those supposedly teaching comprehensive sex-ed considered [abstinence to be] the most important message they were trying to convey, thus bypassing important teaching about contraception or prevention of STDs,"[84] reports writer Peggy Orenstein. She also says that for most teenagers today, the children of those 1990s adults, female pleasure is not a part of their sex education in any way at all. U.S. government funding since 1996 for such abstinence-only-until-marriage programs currently stands at $2.1 billion[85] and it is still going strong. It was not only the 19th century that limited access to information about sex and sexuality.

Georg Ludwig Kobelt may have put the clitoris on the anatomical map in the mid-1800s, but it was a time when anxiety about female sexual pleasure was growing—remember, women were already getting off on the speculum. This was all too much. It was as if woman *had* been given two penises after all. What a cruel trick for nature to play on her favorite son. It really didn't seem a good idea for the masses to know too much about it, given women's vulnerability. A rather sad afterthought is that ultimately perhaps Kobelt's knowledge only encouraged the surgical men who came up with ways to solve the problem of the clitoris by operating on it, rather than educating the next generation about women's anatomy in more joyous ways.

A Note on Female Genital Mutilation

It is estimated that globally three million girls a year are at risk of FGM, a procedure that often includes the partial or total removal of the clitoral glans. This book is about the clitoris in Western culture, and while communities within the Euro-American world participate in FGM, it is as part of their cultures and not the hegemonic Euro-American culture. However, I have included an appendix on FGM because it is far too important an issue to leave out.

Sexually Adequate Women
Do they even like sex?

Our poor clitoris couldn't win. Regardless of whether she was an anomaly or the female equivalent of the penis, it all came down to the same thing: she wasn't welcome. Remember the days when a quick third-party orgasm was the cure for hysteria? The clitoris had gone from being a button for alleviating hysteria to a switch for sabotaging the enlightenment project. Men became increasingly nervous about that. American doctor George Taylor, who had patented steam-powered massage and vibratory apparatuses for female disorders in 1869 and 1872, now warned physicians that treatment of female pelvic complaints with the "Manipulator" should be supervised to prevent over-indulgence. And French physician Henri Scoutetten had this concern in 1843 about the popularity of the cold-water douche with his women patients:

> The first impression produced by the jet of water is painful, but soon the effect of the pressure, the reaction of the organism to the cold, which causes the skin to flush, and the reestablishment of equilibrium all create for many persons so agreeable a sensation that it is necessary to take precautions that they do not go beyond the prescribed time, which is usually four or five minutes.[86]

What did these men think was going to happen if a girl ran a little over time? What exactly was the problem if she came back for more? FFS, give us a break! In 1903 physician E. H. Smith was so anxious about his colleagues failing to diagnose masturbatory diseases in female patients that he wrote and published a guide to detecting them for the *Pacific Medical Journal*. A sure sign was having labia of unequal size. He believed that such "hypertrophy" was caused by masturbation on the longer side, which leads me to believe that not only did he not know much about female genital anatomy, but he also understood very little about how vulva owners self-pleasure.

Another female disorder: frigidity

Toward the end of the 19th century, science found another bogus female illness to add to time-honored hysteria. The previously unacknowledged problem—some women's failure to orgasm during sex—started to be considered a pathological condition: that of being frigid. Until writing this book, I had always assumed that *frigid* meant "cold" or "unenthusiastic about sex"; but by the early 1900s the word very specifically referred to a woman being "unable to orgasm with coitus."[87] It was an infliction that carried shame and judgment. Additionally, some men believed that women who failed to orgasm during intercourse were willfully withholding their orgasms in an act of defiance that was deemed unhealthy, as it signaled either lesbian tendencies or a subversive disrespect for men. Being frigid was an entirely female matter, like hysteria. As if men don't suffer from it. But wait a minute! The week of May 8, 1998, one month after Viagra was launched, 300,000 Viagra prescriptions were written in the United States.[88] What's that about, if not a man's version of being frigid?

Led by Sigmund Freud, a founding father of psychology, the 19th-century medical world began categorizing women's orgasms into vaginal and clitoral orgasms. During this time, psychology (evolving from the anatomical science that had gone before it) became the dominant science behind thinking on female sexuality. Freud provided a solution to the fear about women masturbating by declaring that vaginal orgasms were the right type of orgasm. Clitoral orgasms, on the other hand, were immature and indecent, solitary and masturbatory in style. They were a threat to society! I think he meant they were a threat to men who subconsciously worried that their penises might not be so special after all, and who were frightened to give women control over anything beyond raising small children and organizing housework, and who displaced their own guilt about solitary masturbation onto women. Hours of Freudian psychoanalysis were prescribed for women who preferred clitoral to vaginal stimulation, so that they could be cured of their inappropriate preference.

The successful 1966 sex manual *The Sexually Adequate Female* (reprinted at least 19 times) was marketed as "an indispensable guide to . . . one of the most complex, delicate, confusing, and least understood problems of womankind . . . how to function adequately sexually." At its heart was woman's maladaptation as a sexual being. Frigidity, the reader was told, is the most common sexual "failing" among women. On page 64, the author, Frank S. Caprio, M.D., recommends "psychiatric assistance" for any woman who prefers clitoral stimulation to other forms of sexual activity. Is it any wonder our grandmothers didn't talk about the clitoris?

A surgical solution to the clitoris problem

In an era of medical advancement in which germ theory, anesthesia, and X-rays were developed, the belief in the power of medicine to cure mankind's ills started to extend to the realms of female sexuality as well. Deviation from the vagina-as-penis model of sexual satisfaction became a curable illness, either through psychiatric therapy or with the help of a surgeon's knife. (This book is concerned with the clitoris, but gay men likewise have suffered greatly from medical attempts to "cure" their perceived failure to conform.) Clitoridectomy emerged in Europe and America, not as a religious ritual, but rather as a "minor" surgery to "fix" women who masturbated, were frigid, lesbian, insane, or diagnosed with nymphomania, depression, epilepsy, catalepsy, or hysteria.

Two infamous proponents of clitoral surgery, albeit for different purposes, were Isaac Baker Brown and James Burt. Brown was president of the Medical Society of London in 1865 and pioneered several gynecological surgical techniques at his London Surgical Home for Women, claiming that epilepsy and nervous disorders were caused by "unnatural irritation of the clitoris." His cure, for which he claimed a 70 percent success rate, was described as a "harmless operative procedure," but was actually a very painful operation that cut off the clitoris. He advised fellow surgeons that "[t]he clitoris is freely excised, either by scissors or knife—I always prefer the scissors." He and Lanfranc of Milan were alike: separated by 600 years, but united by the ease with which a clitoris can be snipped off like a piece of excess fat on a joint of meat.[89]

One hundred years later, James Burt of Ohio, who practiced from 1966 into the 1980s, went by the name the Love Surgeon. His specialty was helping men bring their wives to orgasm vaginally by operating on the clitoris. He claimed his procedures were wildly successful. In his 1975 book, *Surgery of Love*, Burt wrote: "Women are structurally inadequate for intercourse. This is a pathological condition amenable by surgery."[90] Surgery often included removing the hood of a patient's clitoris as well as other vulval alterations, with the objective of increasing responsiveness to intercourse. "Structurally inadequate"? By whose agenda? This structural inadequacy was measured solely by a woman's ability to orgasm during intercourse—but if you cease to make intercourse the nadir of sexual satisfaction, there's no inadequacy. If you judge a fish by its ability to climb a tree, it will live its whole life believing it's stupid.

Both Brown and Burt were discredited during their lifetimes, but not before they'd spent years conducting surgery on women and promoting their cures. *The Church Times* of 1866 alerted the clergy to Brown's practices, endorsing him as an "eminent surgeon" and recommending Brown's "little book," which would enable them to "suggest a remedy for some of the most distressing cases of illness which they frequently discover among their parishioners." The newspaper further urged the clergy to do a service to their poorer parishioners by bringing the surgical technique "under the notice of medical men."[91] Evidence of the continued level of acceptance of this form of treatment in Burt's time is that the Blue Cross Blue Shield Association, one of the most recognized names in health insurance both in the United States (insuring more than 106 million Americans in 2020) and around the world (with a presence in 170 countries), covered these procedures under their insurance policies until 1977.[92] It seems that having a clitoris was as dangerous in the 19th and 20th centuries as it was in medieval and Renaissance times.

I recently came across the 1899 *Merck's Manual of the Materia Medica,* "A ready reference pocket book for the Practicing Physician,"[93] which consists of handy lists of symptoms, cures, and drugs. The medications suggested for treating those suffering from "Sexual Excitement" or "Nymphomania" make toxic reading. The list of "anaphrodisiacs" includes: belladonna, conium (hemlock), gelsmium (British woodbine), hyoscyamus (henbane), opium, and stramonium (jimsonweed). On further research I find this heady mix variously affects nerve impulse transmission and heart rhythm or induces dizziness, hallucinations, the sensation of flight, coma, and, in extreme cases, death by poisoning. That's if you didn't get your clit cut off. In his book *The Lover's Tongue,* Mark Morton traces the history of the word *masturbation,* the emergence of synonyms through the 19th century, and a burgeoning of "innumerable slang synonyms ... at least for male masturbation"[94] in the last half of the 20th century. He points out that terms denoting female masturbation "are few in number." Is it any wonder?

Normal women are not sexual beings.

Despite the paranoia about the clitoris, and running parallel to the commitment to the vaginal orgasm, there was a growing belief that "normal" women were wired to seek pregnancy rather than sexual pleasure. This was advocated with increasing frequency during the late 18th and early 19th centuries. In 1886 German psychiatrist Richard von Krafft-Ebing, heralded as the earliest of modern sexologists, published an influential and popular reference book about sexual practices written specifically for doctors, fellow psychiatrists, and judges (men by default). *Psychopathia Sexualis* was translated into English and was so popular that it had run to 12 editions by 1903. He was of the opinion that woman, "when physically and mentally normal and properly educated, has but little sensual desire. If it were otherwise, marriage and family life would be empty words."[95] Despite acknowledging the lack of sexual satisfaction felt by many women of his day, he attributes it to their being wired to want the status of marriage and children, rather than their sexual needs not being met within marriage. Come back De Graaf, who couldn't see why women would put themselves through the irksome business of pregnancy without the incentive of sexual pleasure and passion.

Krafft-Ebing does not even use the term *orgasm* in his book, but writes about pleasurable feeling and pleasurable sensation, saying that the sensation occurs earlier in men than in women, and that in women it outlasts a man's act of ejaculation. So even if women did get to read this book—which was not Krafft-Ebing's intent as he explains in his introduction to the first edition where he says, "In order that the unqualified persons should not become readers the author saw himself compelled to choose a title understood only by the learned"—they would learn nothing about their sexuality beyond an endorsement of the feeling one is left with after non-orgasmic sex! No surprise then that when Dr. Katharine B. Davis started her study of the sex life of some American women in the early 1920s, she found the word *orgasm* "unfamiliar to many."[96]

Estimates from this era as to the percentage of women who took no pleasure from sex concur with what more formal 20th- and 21st-century sex surveys have consistently found about the number of women reporting they do not orgasm regularly as a result of intercourse. A statistic starts to emerge: three-quarters of women do not derive orgasmic pleasure from intercourse alone. A number of 19th-century medical sources claimed "frigidity" rates of between 66 and 70 percent for "civilized" women. In his 1927 *Studies in the Psychology of Sex*, Havelock Ellis assesses the evidence from numerous academics. He quotes leading German gynecologist, Hermann Fehling, who declared in an address at the University of Basel in 1891 that "[i]t is an altogether false idea that a young woman has just as strong an impulse to the opposite sex as a young man," adding that "half of all women are not sexually excitable." From France he gives us Adam Raciborski, who found in 1844 that "three-fourths of women merely endure the approaches of men." Scottish surgeon Lawson Tait reported in 1801 that "women have their sexual appetites far less developed than man." Italians Cesare Lombroso and Guglielmo Ferrero were of the opinion by 1803 that "[w]oman is naturally and organically frigid." In 1883 William Hammond (surgeon general to the United States Army, and a neurologist) wrote in his book *Sexual Impotence in the Male and Female* that "it is doubtful if in one-tenth of the instances of intercourse [women] experience the slightest pleasurable sensation from first to last." This is not the case everywhere, Ellis tells his reader; it was not true of women in antiquity, nor among the vigorous barbarian races of medieval Europe, and is not true for women from Russia, Scandinavia, Peru, or China. In summing up, Ellis quotes a Dr. Harry Campbell: "The sexual instinct in the civilized woman is, I believe, tending to atrophy."[97] In 1895 the gynecologist Robert T. Morris had even written an article, "Is Evolution Trying to Do Away with the Clitoris?" He should have switched out the word *evolution* for *society* and it would have been more accurate. Instead he hypothesized that the clitoris was larger among women in tropical climates while disappearing from "civilized" women, thus proving that evolution was indeed doing away with female sexuality.[98] We are back with the legacy of Saartjie (Sara) Baartman.

What about the belief that orgasm was necessary for conception? Science had done away with this notion too. In 1780 Lazzaro Spallanzani had succeeded in artificially inseminating a water spaniel, implying that female dogs at least could conceive without an orgasm. People were not blind to the reality that women who had clearly not enjoyed their sexual encounters fell pregnant, and there was a growing belief that while orgasm might in some way be related to conception, it was possible that it could occur in women and they might not feel it. Then, in 1827, Karl Ernst Von Baer formally proved the existence of ovum by looking at them under a microscope, and the female reproductive cycle was fully understood. Women released eggs once a month irrespective of what was going on in their sex lives. Young virgins released them, nuns released them, spinsters released them. Who needed an orgasm?

Marital advice manuals reinforced the stereotype of women as not sexually excitable and bolstered the social mores of the time.[99] Emma Frances Angell Drake's 1902 *What a Young Wife Ought to Know* made it clear that the objective of being a wife was to get pregnant, "[o]therwise she has no right or title to wifehood." Walter Gallichan's 1918 *The

Psychology of Marriage urged that the virgin bride should be better prepared for marriage, on account of the "outrage" many experienced on the bridal night. But how were young brides to be prepared when literature had erased their pleasure from texts available to them? Maurice Bigelow's 1916 *Sex-education: A Series of Lectures Concerning Knowledge of Sex in Its Relation to Human Life* contained the advice that the bride should know the scientific names of her reproductive organs, but that the word "vulva" would suffice for her external organs because "[d] etailed description of the external organs might arouse curiosity that leads to exploration and irritation." What a world apart from the marriage advice books of the 16th and 17th centuries such as *The Midwives Book* or *Aristotle's Masterpiece*. Let's hope they also read Marie Stopes's 1918 *Married Love*, which tried to normalize female sexual pleasure, lamented the widespread view that women were "supposed to have no spontaneous sex-impulse," and at least mentioned the clitoris . . . once,[100] although it was banned in the U.S. until 1931 under the obscenity laws.

The notion that women were not sexual beings was erroneously conceived from the fact that most women do not orgasm through intercourse. Most women *are* orgasmic. Current research suggests that these days many women have masturbated at some point, mostly to orgasm, and that under their own steam these women reach orgasm within time frames similar to that of men, i.e. within three to four minutes. But given the range of "cures" offered to women who masturbated or were not taking satisfaction from the marriage bed, one can see why women might either pretend they were having a vaginal orgasm or buy into the idea that they weren't sexual creatures anyway.

Robert Latou Dickinson's *Human Sex Anatomy*, published in 1949, was singular for its attitude. "Exalting vaginal orgasm while decrying clitoris satisfaction is found to beget much frustration. Orgasm is orgasm however achieved." A lone voice shouting against the clitoris-denying throng. As recently as 1977, in *A Synopsis of Gynecology,* Daniel and Woodard Beacham noted that "during stimulation the clitoris fills with blood becoming larger and firmer," but, they continued, though the organ "is richly supported with nerves," it is "not essential for orgasm if the patient has developed a normal sexual reaction pattern." The Beachams had clearly not read *The Hite Report: A Nationwide Study of Female Sexuality,* published in 1976, which asked hundreds of women what they experienced, liked, and did during sex—not to mention the research by Alfred Kinsey, which had started talking about the role of the clit in female sexual pleasure in the 1950s.

Famously, the 1948 editor of the 25th edition of the go-to anatomy textbook, *Gray's Anatomy,* left out the clitoris. True, the 1901 edition only indicated it as a small node, but the 1948 editor, Dr. Charles Mayo Goss, omitted it entirely, despite De Graaf's and Kobelt's clitoral studies. He retained numerous other tiny details of the anatomy, like Gimbernat's ligament. Sadly, he isn't alive to explain why or how the clitoris was missed under his editorship of this flagship work, but it was an extraordinary omission. Did no one notice? Did they notice and think it didn't matter? Or was it left out because it was inflammatory?

In their review of representations of the clitoris in anatomy texts in the 20th century, Lisa Jean Moore and Adele E. Clarke[101] found that despite the famous studies by Kinsey (1953) and Masters and Johnson (1966), demonstrating the importance of the clitoris in women's sexual function, "the absence of a labeled clitoris in the era 1953 to 1971 is especially noteworthy," particularly in the face of the 1970s feminist movement and Shere Hite. Overwhelmingly, they found diagrams of the clitoris since 1981 still to be very simple, generally with only one part of it labeled; and yet texts tended to describe the vagina in full-blown evolutionary functionalist theory, as a receptacle for the penis. Full clitoral anatomy is still not included in many current medical textbooks, whereas there is no shortage of coverage on the penis. There has been more at play than simple misunderstanding, more than just a lack of scientific knowledge. By redirecting women to their vaginas, there has been an unspoken campaign to gaslight the clitoris and the pleasure it can give women. Why can't we have both, like work and parenthood?

Alvin Silverstein's popular 1988 *Human Anatomy and Physiology*[102] lends support to this conspiracy theory. "With the current emphasis on sexual pleasure and the controversy over the role of women (and men) as sex objects, it is often easy to lose sight of the fact that a large part of woman's body is adapted specifically for functions of conceiving, bearing and nurturing children," he says. While acknowledging the clitoris as an important region of sexual stimulation, he goes on to undermine it, saying it has become "something of a cause célèbre for feminists rebelling against the 'myth of the vaginal orgasm.'" By highlighting the phrase "myth of the vaginal orgasm," he suggests this concept was created by feminists, and implies that the vaginal orgasm in his view is not a myth. When he wrote this, in 1988, *so* much evidence already existed to prove that, for many women, an orgasm experienced vaginally *is* a myth. Women weren't calling it a myth to be rebels. They were neither rebels nor frigid: they were telling the truth about their lived sexual experiences.

So how did female sexual pleasure and the clitoris get written out of the script? That has to be one of the driving questions for this book. For many centuries, science misunderstood women's anatomy and remained wedded to both androcentric and patriarchal visions of female sexuality; this gives us our first answer. The most problematic issues with the vagina-as-penis version of woman are that either it denied any sexual function for the clitoris, which became nothing more than a crisis-management solution for hysteria, or it brought about the judgment that a clitoral orgasm was an inappropriate sexual response. Marginalization of the clitoris also ultimately led to the belief that women were naturally less sexual than men, the flip side of this being that men were deemed to be naturally sexual, which gave them moral leeway and sexual license. The big lie has been the prowess endowed upon the penis to provide complete sexual satisfaction. This was so deeply assimilated into European and American culture that any other model of sexuality became inconceivable.

The second answer to the question of how the clitoris was sidelined lies in the aversion to masturbation that long preoccupied society. If we want to close the orgasm gap then people must find the language and courage to show their partners how to stimulate them to orgasm, or indeed, just get on with it themselves as part of an encounter that includes intercourse. It means overcoming your coyness about masturbation and yes, sometimes doing the work yourself. For some men, it means getting over their penises, getting over being judgmental or weird about women masturbating, and also putting in some work. Research data on women orgasming as part of a heterosexual encounter shows that it is, unsurprisingly, more likely to happen with established couples or for women who are comfortable using the word "clitoris." Just under half of heterosexual encounters that end in orgasm for men don't result in an orgasm for the woman, while lesbian encounters (and presumably those with FTM trans partners) report much higher orgasmic satisfaction. Why should cis men know what to do to achieve a more equal orgasmic outcome when the clitoris is so invisible and female masturbation equally taboo?

We are patching together a history of the clitoris that almost entirely lacks any female perspective beyond Jane Sharp's account in the 1600s. How did women talk about the more intimate details of their lives? What secrets were shared among women as they engaged in women's work, or withdrew to drawing rooms after dinner? What advice was passed on by grandmothers, mothers, aunts, sisters, and friends about what it was to be a woman, about sexual pleasure and sexual intercourse? We don't have a written record, although my guess is that each family or friendship group had its own subculture, just as there is today, and that within these groups the information, misinformation, and cultural expectations played out differently. There is a gap in the historical record, just as there is an orgasm gap.

2
The Truth About the Clitoris

How big did you say she was?

An argument ensued, each blind man thinking his own perception of the elephant was the correct one. The rajah, awakened by the commotion, called out from the balcony. "The elephant is a big animal," he said. "Each man touched only one part. You must put all the parts together to find out what an elephant is like."

I remember seeing two versions of woman's genital organs in biology and sex-education lessons. The first, the ram's-head diagram, most frequently used for learning about reproduction, periods, and sex, is a vertical cross section through the middle of a woman's body.

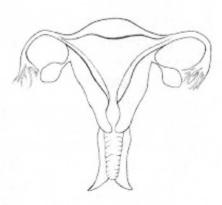

The vagina is visible and shown in a format that enables one to understand how a penis might fit in. The Fallopian tubes curve around toward the ovaries at the top, creating the ram-like look of the diagram. The uterus is central. The clitoris is not featured.

The second familiar diagram is the one that shows a woman's legs splayed. This is, if you like, the horizontal plane, with the woman on her back, and is the angle frequently associated in our culture with sexual intercourse and birth. It

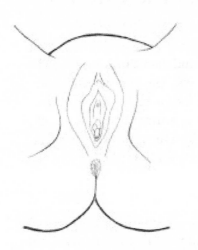

presents concentric ovals to indicate the mons pubis and labia, and a couple of circles for the vagina and anus. The urethra and clitoris are shown at the top, usually as small dots; that is, if you are lucky and get a clitoris at all. Diagrams like this are used as graphics in birth explanations and books for girls about their own anatomy, when they are sometimes accompanied by a suggestion that you sit naked with your legs apart and use a mirror to view your own version. Neither is helpful or accurate in terms of the clitoris.

The true anatomical extent of the clitoris was definitively mapped in 2005 by Professor Helen O'Connell with an MRI scanner. Although, as we know, anatomists had discovered much of this information previously, O'Connell's team provided irrefutable evidence. O'Connell, a urologist, first published her work in 1998 and since then, combined with the advent of MRI scanning and 3D printing, it has changed forever what we know about the clitoris. She confirmed that the small external bud of the clitoris seen on diagrams—and in a hand-held mirror—is the tip of the clitoral iceberg. The clitoris extends internally to include a pair of long arms, typically 3½ inches long, and a pair of fleshy bulbs that are fuller and reach about 2¾ inches in length. A whole clitoris looks like this:

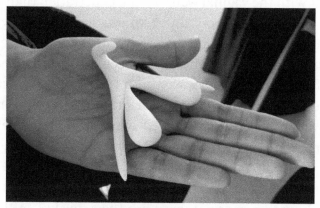

3D-printed clitoris, to scale.

This is how it fits with the rest of a woman's genitalia:

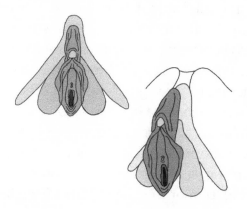

Clitoral anatomy, showing vulva and internal structures.

The clitoris is not small: it exists on a number of planes, and partially surrounds the vagina. It is made of erectile tissue that swells when stimulated. This erection occurs in both the visible external nub of the clitoris and also throughout the internal body. Erect bulbs are far more likely to come into contact with the vaginal walls during penetration than flaccid ones. During sexual arousal and stimulation, blood flow in the vaginal walls increases approximately threefold. Blood flow into the clitoris, on the other hand, has been estimated to increase from four to eleven times baseline.[103] This vasocongestion is released by the pelvic contractions that occur with orgasm. Yes, women get erect clitorises, feel physical sexual desire in the same way that men do, and when aroused can feel the same frustration when the release of orgasm does not occur.

Like bygone anatomists, O'Connell and her team's initial research was on cadavers, but with the arrival of MRI scans, living women provided the most accurate understanding of the clitoris. The clitoris is responsible for many orgasms experienced through genital stimulation; it's just a matter of where the stimulation occurs. For some, vaginal stimulation of the clitoris brings them to orgasm—either through the rubbing of the glans during intercourse or through the proximity of the internal structure to the vaginal wall. For others, having the glans stimulated so that the internal bulbs and crura are fully aroused enables an orgasm to occur with coitus. For many people, however, no amount of stimulation through the vagina is sufficient to trigger an orgasm—it can only be experienced through stimulation of the glans. That's just the way some bodies are made. Not orgasming in intercourse is not a "disorder" or a "failure"; it can be a function of physiology. While it is true that some women can learn to orgasm through intercourse—by means of practice, knowing what they are aiming for, finding new positions, asking for the right foreplay, or small adjustments that make a difference—research also shows that for many women, unaided coital orgasms will always remain elusive.[104]

In the past decade, scientists have tried to pinpoint the exact factors that make the orgasmic difference. A 2011 research study[105] reported that shorter external distances between the glans clitoris and the urethral meatus resulted in better orgasms during intercourse. A more recent project[106] using MRI measurements showed that women with coital orgasmic function had a larger clitoral glans and a shorter distance from the vaginal lumen to the clitoris. In 2008, ultrasounds demonstrated that women in a sample who were able to accomplish "vaginal orgasms" had a thicker ureterovaginal space, which indicated that clitoral mass in that area is key.[107]

To date, sample sizes have been small, and I would hate for this kind of research to throw up another measure by which women are encouraged to gauge themselves, along with how their labia should look. I recently drove behind a bus that advertised "Vaginal Rejuvenation, Labiaplasty, Mons Pubis Reduction, and Labial Puffs." It would have been funny if it was a Monty Python sketch, but sadly it was for real: the "most intimate makeover," on offer from a cosmetic gynecologist.

Recent research by Dr. Beverly Whipple in her human physiology laboratory—and in functional magnetic resonance imaging (fMRI) of the brain during orgasms—found that women can experience orgasm from imagery alone, and women with complete spinal-cord injury can also experience orgasm from vaginal and cervical stimulation.[108] Orgasm is experienced in the brain from a number of nerve pathways, and the vagus nerves bypass the spinal cord. Some women experience orgasm from non-genital stimulation of a body part, for example a nipple or the mouth. And yes, some women experience orgasmic dreams. I would love to know whether the clitoris fills with blood to mirror the brain activity, or whether the brain alone can give orgasmic sensation. Can some women orgasm through vaginal and cervical stimulation without any arousal in the clitoris, or is it indirectly engaged? Female sexual pleasure is much more nuanced than body measurements.

The question of female sexual pleasure in the older woman seems a niche interest, and it was hard to uncover research on the clitoris and age. I have a British friend who recounts her quest to find a gynecologist when she first arrived in the U.S. Having a pap smear or being prescribed hormone replacement therapy in the U.K. is a matter for your general practitioner. Of a certain age, and facing the tedious effects of being post-menopausal, my friend was excited by the prospect of access to a specialist dedicated to her vulva. In her forthright manner, she related her search as if she was performing an Alan Bennett monologue.

"So, there I am in my gown. Once we'd done the form-filling and pleasantries and she'd palpated my breasts and peered at my cervix, I started telling her about sex. I said, 'It's just not the same . . . it takes more work, and orgasms aren't the way they used to be. All that effort for a fizzle rather than glorious fireworks.' She said, can you believe it, 'When you get to this age, cuddling and affection become more important.' I thought, fuck that for a game of soldiers, I've done enough cuddling and dishing out of affection with my children to last a lifetime. I want to know, where's my orgasm? So much for finding a gyne-cologist my age, thinking she'd be sympathetic."

I nod to show I'm listening and open my mouth to comment supportively, but my friend holds up a finger, paus-ing my response, and continues.

"I went to a second gynecologist, and after the pleasantries, palp, and peer, I said the same thing, about the fizzle rather than the fireworks, and you know what she said? 'Firstly, you are very lucky that you have a husband; secondly, you are very lucky that he can still get it up; and thirdly, you are very lucky that he still wants to get it up with you.' More fucking soldiers, I thought. And I looked her in the eye and said, 'He is very lucky to be getting it up with me! And you're supposed to be on my side.'"

My friend takes a breath as I nod again, realizing that her monologue is not over.

"This is the problem for older women these days. We're fitter, more groomed, and hotter than ever before—I mean, look at me!—and not worried about pregnancy, and bang! Sexually, you're done. Cheated. Cheated by nature and cheated by the system. Even gynecologists don't want to know! I went to a third gynecologist. She kept me waiting, but instead of getting into the blue gown, I looked around her consulting room. This time, when she came in, I said, 'Before we do the form-filling, pleasantries, palp, and peer, are you willing to talk about sex, specifically my orgasms? I know there are pressing issues in gynecology, like cancer and fertility, but for this half hour, I am your patient, and I want to talk about sex. And by the way, your model of the female genitalia, you know, it's missing the clitoris.' Know what she said? 'I'd never noticed.' She must have seen how riled I was, because she very quickly added, 'But I am very happy to talk about sex. Let's start with understanding what your hormones are doing.' Thank God! A little bit of progesterone later and you know what? Fireworks are back. Are they the same? It's not New Year's Eve yet, but it's fireworks. I'll take them." She subsequently told me that pelvic-floor PT had worked wonders in bringing the display back to its fullest glory.

The popular press frequently refers to a locus of woman's sexual function that hasn't been discussed yet: the G-spot. It's named after Dr. Ernst Gräfenberg, who first described "an erotic zone located on the anterior wall of the vagina along the course of the urethra that would swell during sexual stimulation" in 1950. At the time, no one took much notice, but sex scientists, led by Dr. Beverly Whipple, returned to his work in the 1980s when seeking the key to female sexual pleasure was more in vogue. The G-spot became a thing and people set to work to find it.

Dissection on cadavers and MRI scans have not revealed any additional structure that is separate from the clitoris, urethra, or vagina that could be regarded as the G-spot, but it is, as Dr. Whipple explained to me, "the female prostate gland [and] the bulbs of the clitoris, as well as other tissues in the area." French gynecologists Odile Buisson and Pierre Foldès did studies using ultrasound that looked at the stimulated clitoris and the clitoris during vaginal penetration.[109] "The G-spot," they noted in a 2009 paper, "could be explained by the richly innervated clitoris." This means that for some people, the aroused bulbs of the clitoris press around the vagina and create a small dip in the anterior wall that is highly sensitive to touch, because the clitoris is so close to the vagina wall. It makes sense that if the internal body of the clitoris is anatomically situated close to the vaginal walls within the pelvic cavity then one would be able to orgasm through this form of stimulation. Latterly Emmanuele Jannini has provided data to support the anatomical relationships and dynamic interrelations between the clitoris, urethra, and anterior vaginal wall, and labeled the region of the Gräfenberg spot the *clitourethrovaginal* (CUV) *complex*.[110] The internal shape, size, and location of a person's clitoris is as variable as any other part of their anatomy and the clitoris's interconnectedness within the pelvic region is so physiologically intricate that we are only just beginning to understand it. Women's orgasms and clitorises are as diverse and glorious as their owners.

The implications of having full anatomical knowledge of the clitoris go much further than settling any Freudian debate about the nature of women's orgasms and whether they are experienced vaginally or through the clitoris: orgasms involving genital stimulation are clitoral. The question is, are they triggered through stimulation of the glans or the whole? and this depends on individual anatomy. Knowing this, people can understand their own sexual experiences. It also allows us to reset expectations around orgasm and how it is experienced, giving many people permission to stop searching for the holy grail of the vaginal orgasm and allowing them to enjoy the stimulation that works for them. It frees up men from feeling that the onus for a woman's orgasm is on the penis. It makes conversations about what works sexually for one's partner(s) central to mutually orgasmic sex.

Accurate anatomical knowledge also impacts on pelvic surgery. Pierre Foldès, a surgeon, used it to pioneer reconstructive surgery for women who have experienced female genital mutilation (FGM). In 2012, he and his team reported in *The Lancet* that over 11 years, they had operated on nearly 3,000 women. At one-year follow-ups, attended by 866 of these patients, 815 reported clitoral pleasure and 431 experienced orgasms.[III] It should be noted that this work is controversial and lacks a control group. In a letter to *The Lancet,* a heavyweight British team—consultants in gynecology, obstetrics, and psychology—took issue with the study, saying, "The claims are not anatomically possible ... Where the body of the clitoris has been removed, the neurovascular bundle cannot be preserved ... There is therefore no reality to the claim that surgery can excavate and expose buried tissue." However, other surgeons continue to report success with Foldès's procedures. For example, according to Dr. Marci Bowers, an obstetrician and gynecologist in California interviewed by Global Woman P.E.A.C.E. Foundation in 2016:

When restoration is performed, the clitoris is found 100 percent of the time. True, there may be damage to the tip, but the tip is, literally, the tip of the iceberg. Restoration is possible by not only dividing the infibulation (if present) but clearing the remaining clitoris of scar tissue and securing it to the surface skin. The sensory clitoral majority remains after FGM— fortunately, thanks to Dr. Pierre Foldès, there is now a technique to access those portions.

Also, for trans men wishing to go beyond metoidioplasty— which works with an existing clitoris to form a neo-penis—to phalloplasty surgery, understanding of the clitoral anatomy enables its incorporation into a newly formed phallus to hope- fully achieve a site capable of orgasmic pleasure.

There is no excuse for mainstream sex-education books to leave out the full anatomy of the clitoris. Much sex that occurs today is not for reproductive purposes. If you're teaching about contraception, why are you not also teaching about mutual pleasure? Unless you're concerned only with procreative sex, leaving out the clitoris or diminishing it in any sex-education forum is no different from leaving out the penis or indicating it only as a small diagrammatic slash.

Are you responsible for a teen? Maybe it's time to get to grips with the sex-education curriculum at their school and understand how it sits with your values about healthy and happy sex. If you're not having conversations with your teen now about how sex works, it's unlikely to be a topic that you will be returning to, or able to build on in the future, so what they learn there is crucial. Does the school follow a liberal program, espousing the idea that moral sex is consensual sex in which neither party is harmed, and believing that children need information so that they become empowered to make good choices for themselves? Even if it's liberal in outlook, does female pleasure make it onto the agenda? Or does the school follow a more conservative model, seeing itself as the gatekeeper to children's morals and protector of their innocence, with an emphasis on safe reproductive sex and possibly, encouraged by federal funding, promoting abstinence? Within this latter framework, there is unlikely to be room for a discussion about the clitoris and clitoral pleasure. Let's hope for more diagrams like these in future:

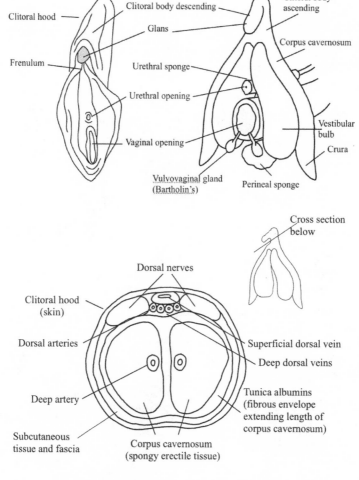

Clitoral details, including cross section.

Scientists don't operate in a vacuum; they work within a cultural context, and it is naive to deny the power of a culture to shape the academic work that it sponsors and champions. Engagement with the anatomy of women and female sexual pleasure has not been purely down to anatomists and scientists. Attitudes have been shaped by other cultural drivers, like religion and philosophy, and their role in the evisceration of the clit is arguably as big.

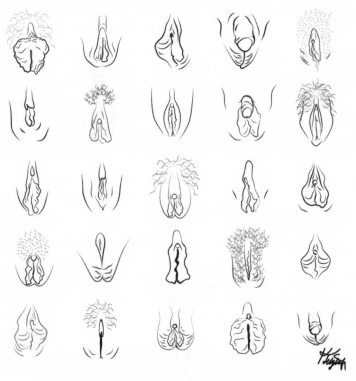

Clitorises are as diverse and glorious as their owners.
(Medical drawings by Katja Tetzlaff.)

PART TWO:
PERCEPTION

3
Female Immorality and Sexuality
Why do they go hand in hand?

> There was only one catch and that was Catch-22, which
> specified that a concern for one's safety in the face of
> dangers that were real and immediate was the process of
> a rational mind. Orr was crazy and could be grounded.
> All he had to do was ask; and as soon as he did, he
> would no longer be crazy and would have to fly more
> missions ... Yossarian was moved very deeply by the
> absolute simplicity of this clause of Catch-22 and let out
> a respectful whistle.
>
> "That's some catch, that Catch-22," he observed.
> "It's the best there is," Doc Daneeka agreed.
>
> Joseph Heller, *Catch-22*

One damaging idea that impacted our anatomical knowledge about the clitoris and our understanding of female sexuality was that woman was a version of man. But there is a second dangerous concept that repeats in various hues in the patchwork history of female sexuality: the persistent notion that women are morally as well as sexually lax, and that one is symptomatic of the other. It is a chicken-and-egg paradox, and it's hard to unravel which came first. Why have the two been so inextricably and unfairly linked? This symbiotic relationship exists only for women and is still played out today.

Imagine the shock if a female president was discovered to have enjoyed cunnilingus from an intern at her desk in the Oval Office, and you will begin to see how the gendered one-sidedness of this way of thinking has been prevalent throughout history. I made a list of American presidents who are known to have had extramarital affairs and came up with twelve. That doesn't include those who were not married but had illegitimate children (add these in, and there are many more than twelve). By my count, at least a third were not held responsible or accountable for their sexuality in the way that women of the time were. As for British monarchs, Charles II, James II, and William IV had more than ten illegitimate children each, and that's just the ones they acknowledged. Unlike women, men's sexuality has not been regarded as symptomatic of their moral propriety. And don't give me Catherine the Great as a counter-argument and proof that women get the same leeway. She was an exception rather than the rule, and most of those stories were made up to discredit her . . . because she was a powerful woman. No wonder Elizabeth I took the Virgin Queen route.

How did views about female morality come to be bound up with women's sexuality? And where did the belief that women are less morally able than men come from? Answer these questions and we will go some way to understanding why the history of the clitoris in the Western world is complicated and conflicted, not only in the arena of science.

The Church
What good can come from the clitoris?

The trouble with Eve

Some people argue that Christianity, whose key female didn't even have sex, has been instrumental in establishing a culture that labels women as sexually and morally lax. But before the Virgin Mary there was Old Testament Eve, who predates Mary by five or six centuries. What exactly did Adam and Eve learn from biting the apple that grew on the tree of knowledge? Does their modesty with the fig leaf imply that their newfound knowledge had something to do with their genitals? Is the gesture symbolic of a sexual awakening, or the advent of lust in their lives?

We know the serpent was Satan in disguise, but how was Eve to know? All she had been told was not to eat an apple from the tree in the middle of the garden. How many of us hear an instruction but don't take it as sacrosanct until we've learned the lesson the hard way, by disobeying it? How many of us have failed to take a coat at our mother's behest and have then been frozen late at night at a bus stop; or have drunk too much alcohol despite being told that mixing wine and vodka never ends well? Or have—maybe just once—made a poor decision when it came to an attractive admirer? I bet Eve would have been warier second time around. All Eve saw was a handsome snake, lascivious in his voice and slick with his come-on patter, as he slowly slid down the tree and invited her to take a bite of an apple, promising her she would become god-like if she did so. Eve darling, if it sounds too good to be true, it probably *is* too good to be true, didn't your mother tell you? Would you have listened if she had?

Eve was tempted and didn't resist the serpent's invitation. He probably had the voice of Jeff Goldblum or Idris Elba.

Standing under the tree of knowledge, Eve stretched up, cupped the ripe apple, and it dropped readily into her hand. She parted her lips, opened her mouth, and reached her tongue down over the smooth, ripe skin to take a small bite. She curled the crisp wedge back into her mouth, savoring the tart juice on the sides of her tongue. The taste rounded out with a sweet note as Eve bit into the pale flesh, feeling its firmness before she heard the crunch. Eve raised her eyes expectantly to meet those of the large, muscular, unfurling snake, her senses alert to the apple's promise, and her mind thrilled by the prospect of becoming god-like.

The narrative is about the promise of knowledge and power, both of which—as well as the snake—are sexy. If you ask me, it's a heady melting pot of sexy, and Eve fell for it, shared it with her man, and everyone was cursed. Eve was cursed with increased pain in childbirth *and* subjugation to Adam, and Adam was cursed because he followed Eve's lead, thus setting in motion a tradition that men shouldn't listen to women, along with the idea that male dominance is a given. Can a disrespect for female judgment, female morality, and female sexuality be traced back to the Genesis story? Is this where the idea originated that no good can come to a man from a knowledge-seeking, sexual woman? Eve taught Adam, and once Adam knew, he couldn't unknow, poor guy. His sexuality was not his fault … but Eve, she needed to make reparations and be kept under control. If she'd had a modicum of self-control or sense of right from wrong in the first place, none of this fall from grace would have happened and man could be enjoying eternal life in the Garden of Eden, none the wiser, happily sipping a mocktail adorned with a little umbrella and an innocent cherry, instead of slogging his guts out to make a living. Never trust a woman when it comes to sex or moral standards. Men? Ah, well, that's different.

Hang on a minute—what gender was the snake?

As an aside, I'm struck by how well Eve would do in today's world, where colleges, internships, and employers are all looking for candidates who demonstrate intellectual curiosity. She'd be on a shortlist of two, along with Pandora. However, in terms of understanding the cultural landscape that has minimized female pleasure, it seems that Eve played her part by giving in to temptation, and desire became a sign of moral laxity in women. And the biggest temptation in the human adult world? Sex. But wait! Was Eve even real? Or is she a religious construct?

The pamphlet wars: Eve becomes the pattern for all women.

I am not alone in interpreting the Adam and Eve story as having sexual connotations. The ancient Hebrew philosopher Philo of Alexandria stated in *Questions and Answers on Genesis* that the serpent was inclined toward, and a symbol for, passion. "And by passion is meant sensual pleasure."[112] During the 16th and 17th centuries, Europe's so-called *querelle des femmes* raged, focusing on Eve, the nature of women, and their sexuality. Consumed by literate members of the gentry and mercantile classes, pamphlets rolled off the printing presses like British red-top tabloid newspapers in their heyday as writers battled it out, drawing on weighty sources: the Bible had divine status as the word of God; classical antiquity was revered for its wisdom and longevity; anecdotal evidence was used to prove arguments about women; and every rhetorical fallacy you can name was presented to persuade readers that women were lusty and deceitful to their core, like Eve.

The wars kicked off in 1541 with *The Schoolhouse of Women* (reprinted four times by 1572), which bemoaned the sexual appetite of women, alleging that one cock could serve fifteen hens, but fifteen men couldn't satisfy one woman. The author was viciously ribald and inappropriate about women and "the tickle in their tail," imagining Lot's daughters meddling with their sleeping father to satiate their sexual appetites. It was Fake News, but it was news that sold.

For the next 100 years, exciting pamphlets slandering the sexy nature of women sallied back and forth with worthy ones defending female morals. In *The Deceit of Women* the author laments the manner in which Adam, "the most wisest and fairest man of the world," was "deceived" by his wife. Eve is used as a precursor to tales of other lustful and deceitful wives, implying that all women follow Eve in their natures because she is the ultimate mother of all women, which is biblically true. (Erm, the mother of all men too, if you are being this literal. What's that you say about the genealogy of men? Oh, men are the sons of Adam? Of course. How could we forget?)

The most famous pamphlet in this vein was Thomas Tell-Troth's (aka Joseph Swetnam's) *The Arraignment of Lewd, Idle, Forward and Unconstant Women* (1615), which was so popular over the next 20-plus years that it ran to 10 printings. Swetnam's opinion of women was that they were loose, "easily wooed and soon won, got with an apple." I thought the snake tried pretty hard myself, and what he offered was not just any old apple, but an apple from *the* tree of knowledge. However, Swetnam has Eve easily won, which would support his view of women as forward and lewd. It just wasn't a woman's place to want knowledge or power, even if she planned to share it with her husband. Was there some projection going on here? Remember, the dominant scientific knowledge held that women were anatomically inside-out men, and assumptions about female arousal and desire stemmed from this belief. Could it have been that lusty men, given more opportunity and leeway by society when it came to sexual relations, were frightened that women might behave like them if given the chance?

This reminds me of a famous survey, conducted by Russell

D. Clark and Elaine Hatfield in 1989[113] on a university campus, that looked at whether men or women were more likely to engage in casual sex. The research studied receptivity to one specific and unusual sexual offer: a stranger (an attractive stooge) walks up to a subject on a college campus, and propositions sex. All the women in the study declined the offer; most of the men accepted. I will return to a variation on this experiment later, but when it was published it was said to illustrate the easy, uncomplicated sexual availability of men. Yet history has painted women as more sexually willing than men, and this has had implications for the clitoris. It doesn't really matter whether women were lewder or not—what matters is that it was seen as a thoroughly undesirable trait in a woman.

Add to lewd and lusty the accusation that women are untrustworthy. Swetnam blames Eve, and by association all women, for the fall of man. "She was no sooner made," he writes, "but straightaway her mind was set upon mischief, for by her aspiring mind and wanton will she quickly procured man's fall." Swetnam warns men that female "wantons" (sexually unrestrained or having many casual sexual relationships) will enfeeble their bodies with diseases, scandalize their good names, and most seriously of all, endanger their souls. Men, who were blessed by society with intelligence, reason, and fitness for public office, and had authority over women, could be undone by one. Swetnam did not publish as a serious academic; there is a tongue-in-cheek, in-joke element to his pamphlet. It was lowbrow, cheap tittle-tattle—but readers loved it. These jibes reflect the boundaries of the culture for which it was produced and they are an indication of what was fair game. A voracious sexual appetite, laziness, and an inability to remain faithful were stock traits of women in this cultural forum. The clitoris, the seat of a woman's sexual pleasure, would do well to hide herself in this environment.

There were writers who championed women, citing modest, pious, and obedient examples, but their voices held less sway. The most famous pamphlet, and the only one that we know for sure was authored by a woman, was Rachel Speght's *A Muzzle for Melastomus, the Cynical Baiter of and Foul-Mouthed Barker Against Eve's Sex*. Those seeking to vindicate women used counterargument, logic, example, and wit to refute the slanderous, licentious accusations against women. However, these pamphlets were reprinted less often,[114] showing that lewd, bawdy, crude, rude, sensational scaremongering was more popular. The political climate of the time was one where heresy against the Church and rebellion against one's monarch had truly dire consequences. There were bigger battles for men and women to fight and the *querelle des femmes* was nothing more than entertainment. You wouldn't be imprisoned or hung for printing, purchasing, or distributing these pamphlets. They were just a bit of fun, like sexist banter. LOL.

Sadly, few popular pamphlets debated the nature of men. Esther Sournam, whose name is probably both a pseudonym and a pun on Swetnam's surname, lamented, "So in all offenses, those which men commit are made light and as nothing, slighted over, but those which women do commit, those are made grievous and shameful." This is the bind for our clitoris. The territory that was battled over was the nature of women, and that of men was not up for discussion. If I'd had my 21st-century feminist way, I would have argued that some women may be inconstant, with racy sexual appetites, but no more so than some men. As it was, the counterargument tended to go down the route of holding up women as loyal, chaste, and hardworking, marrying "more for the propagation of children than for any carnal delight or pleasure they had to accompany with men, content to be joyned in matrimony with a greater desire for children than Husbands." Furthermore, they found "more joy in being mothers than in being wives."[115] This is a narrative about women that was re-spun by Jean-Jacques Rousseau in the 18th century when he wrote about the ideal woman, and again, as we have seen, by 19th- and early-20th-century doctors and pundits who couldn't understand—or didn't want to endorse—female sexuality.

All female traits, having been tied together—as with personality politics, where one trait is taken to be representative of the whole—stood or fell together. To avoid the banter, bullying, and accusations leveled at them in the course of their daily lives, women needed to model exemplary behavior that contradicted the stereotype being propagated. It was a classic catch-22. It was also a tactic used by some women in America who campaigned for the vote. By proving their moral propriety, they proved their fitness to vote. For women, once sexuality was associated with morality, there was no separating the two. Given that there were huge social advantages in being seen to be moral, female sexuality, with the clitoris at its core, was damned and forsaken.

No number of good women after Eve—even an exemplary virgin—could make up for womankind's sins. The very notion of the Virgin Mary's immaculate conception puts something of a damper on any discussion about sex; and the implied purity of her motherhood over anyone else's is intimidating. Her elevated status within the Church clearly prioritizes chastity. The Virgin Mary and Eve paved the way for the polarization of female sexuality: the predatory, good-for-nothing, lusty temptress (presumably in touch with her clitoris) and the chaste, modest, honorable good girl who would be best advised to hold on to her virginity until she was married (and not to acknowledge her clitoris too enthusiastically). This dichotomy lasted for a long time. Men were advised to seek the latter type for a wife. In an age when women could not inherit wealth, go out to work in the same environments as men, access education, or live or travel on their own without fear, they were on the whole expected to be chaste, and then chastely married—and if you were pragmatic, you had no other option than to conform.

The cult of chastity and chaste marriage
The Catholic Church prizes celibacy. Pope Gregory the Great (590–604 CE) said that all sexual desire is sinful in itself, and priests have been celibate since the 1139 Second Lateran Council approved a rule forbidding a priest to marry. The Council of Trent in 1563 reaffirmed this, stating that celibacy and virginity were superior to marriage. Why have sexual activity and pleasure been perceived as *the* urges that define spirituality and fitness to be close to God and doing God's work in the Christian faith? And why have women been blamed as the temptresses? Don't men take some responsibility for being tempted? "Nothing is so powerful in drawing the spirit of a man downwards as the caresses of a woman," wrote St. Augustine in 401 CE. How about, nothing is so powerful in the spirit of many men as the desire to caress a woman?

In her sophomore year at what was in every other way a wonderfully supportive and caring Catholic high school, my daughter was given a JUG (a penalty or detention, with JUG supposedly standing for Justice Under God) for an incident in a science class. Unfortunately, as she bent forward to light a Bunsen burner, her long-sleeved, collared white shirt, buttoned up to just below her collarbone, which had passed muster through her freshman year as chaste enough, fell forward to reveal her breasts contained in a bra. The reason for her JUG? Distracting the boys. This is an extract from the letter she wrote to her teacher as a result: "I find the notion that my day-to-day clothing automatically makes me the active party in distraction hard to accept. I would like the responsibility of distraction to lie with the parties who allow themselves to be distracted." The teacher who issued the JUG avoided comment by suggesting my daughter could raise the issue with the dean. My daughter decided that her priority was to pass her science class; she didn't want the diversion of a run-in with the male dean about her breasts, or the embarrassment of her mother having this run-in on her behalf. The impossible paradox for woman is that she has been the temptress and yet is encouraged to be sexless, while men have been able to populate the middle ground, protected for many years by a separation between sex and morality.

Apart from making chastity the spiritual zenith, the Catholic Church sees all things sexual as undesirable, excluding the procreation of children, and thus is in sync with early anatomists in terms of reinforcing the penis–vagina sex act as the only one that counts. This is important context for the clitoris because, as the dominant religion in the Western world, the Church has been influential in providing thinking and guidance on female sexuality. This reasoning may have less sway today, but it is a formative part of where we find ourselves.

In 1930, in his papal encyclical (a round-robin letter to Catholic clergy) titled *Casti Connubii*, Latin for "chaste wedlock," Pope Pius XI penned some thoughts on sexual intercourse and marriage. "The conjugal act," he said, "is destined primarily by nature for the begetting of children, those who in exercising it, deliberately frustrating its natural power and purpose, sin against nature and commit a deed which is disgraceful and intrinsically vicious." Readers are left in no doubt as to the status of sex for pleasure and the impact it will have on their relationship with God. Marriages, which should be chaste—meaning conducted morally in the eyes of God—are undoubtedly made base and vile through joyous (but not procreative) sex.

In case anyone was hopeful that the Holy See might revise its teaching, the next pope, Pius XII, made it clear in his 1951 *Acta Apostolicae Sedis* that "[t]his precept is in force today, as it was in the past, and so it will be in the future also, and always, because it is not a simple human whim, but the expression of a natural and divine law." That's a no, then. Given that a man's orgasm in coitus is required for fulfilment of the procreative mission and that 95 percent of men report always or usually having an orgasm with coitus, one half of the population of married, heterosexual couples—namely men—are by default orgasmically happy under God.[116] As are some, but by no means the majority of, women. Elisabeth Lloyd's research suggests that a quarter of women reliably experience orgasm through coitus alone, and this proportion seems to remain constant.[117]

Historian Ciara Meehan reviewed letters received by the agony aunts (advice columnists) of Irish women's magazines in the 1960s. She found that they revealed a shocking lack of accurate knowledge about sex and conception, and also that many women struggled with a "guilt complex" when it came to sex, stemming from the Catholic Church's and the state's view of the female body, which sought to impose on women "standards of idealised conduct."[118] Irish girls were taught from an early age that modesty and purity were valued traits; once they'd internalized these teachings, their letters to women's magazines revealed how hard it was for some wives to switch to obliging their husband's physical desires then reconcile their sexual selves with their moral selves. A Dr. Kennedy, writing for *Women's Choice* at the time, blamed Catholic teaching for women being conditioned to think that it was improper to openly enjoy sex.

Contemporary research on students who have been part

of abstinence programs reveals that more than 80 percent of pledgers forgo their pledge but are woefully ignorant about contraception or STDs. Despite delaying P-in-V intercourse, once they do become sexually active they quickly catch up with their peers in unplanned pregnancy and STIs. Also, heterosexual male pledgers are four times more likely to have had anal sex than their non-pledging peers.[119] One in three girls aged 15 to 17 says she has performed oral sex on a partner to avoid having intercourse, and the same percent say they include oral sex in their definition of "abstinence."[120] All of which should make us question society's definition of what constitutes sexual activity.

I thought I might find a softening in the United States Conference of Catholic Bishops, which acknowledges the unitive power of intercourse, in conjunction with its procreative thrust—but once again the Genesis story of original sin raises its head as we are told that Eve's dalliance with the snake, and her bite of the forbidden fruit, changed the Creator's own gift of innocent coupledom in the Garden of Eden into a relationship of domination and lust.[121] That'll be on you then, Eve, the lust bit. It's all your fault. (Shhh, my love, you are a figment of the male imagination and the victim of a version of events told and retold by men.)

We are further informed by the bishops that "[c]onjugal love is diminished whenever the union of a husband and wife is reduced to a means of self-gratification. The procreative capacity of male and female is dehumanized, reduced to a kind of internal biological technology that one masters and controls just like any other technology."

For the many women who don't experience orgasm through the procreative coitus model, is there room in the teaching of the Catholic Church for female sexual pleasure experienced another way? Or would that be framed as self-gratification? Is the clitoris, like contraception, a bit of biological technology, or can it have a more central role in the sexual dimension of marriage? Despite the relatively low percentage of women who orgasm through coitus alone, a 2015 *Cosmopolitan* survey about intercourse and orgasm found that 57 percent of women usually or always have orgasms during sex. Note the words "during sex."

With a little help, many women are finding orgasmic sexual experiences through oral sex or clitoral stimulation. However, the *Family Life* booklet that came home with my youngest son recently from his Catholic grade school, in conjunction with "*the* talk" that was delivered at school, did not show the clitoris in the diagrams explaining reproduction. Why should it? It's not functional in reproduction. Was the omission in my son's booklet an oversight? A misunderstanding among the largely male hierarchy of the Catholic Church as to how female sexual pleasure works, rather than a denial? At this point, I am reminded of the findings in my survey of sex-education books, discussed later, where there appeared to be no religious agenda and the clitoris was *still* left out.

The historical insistence on procreation as the sole purpose of sex shrouded intercourse in shame and guilt, and possibly encouraged quick, productive thrusting, lest one be getting too much pleasure from the event. This is likely to lessen a woman's chances of an orgasmic experience, even if it *is* something she can experience in coitus. Perhaps some men, who were told that their good wives would find no pleasure in the activity, thought to get it over with quickly. The 2015 *Cosmopolitan* survey also found that the two obstacles to women having an orgasm during a heterosexual sexual encounter were a) not getting enough clitoral stimulation and b) not getting the right clitoral stimulation. The *Cosmo* results are the same as those cited by Shere Hite in her 1976 report on female sexuality. For me, sex without an orgasm is like eating chocolate or a peach without being able to taste it. Quite nice to begin with and, after a while, disappointing. Hite's report also revealed that women who had never experienced orgasm were happier with their sex lives than orgasmic women who were not experiencing orgasm during their partnered sex. This latter (and larger) group held residual feelings of betrayal, anger, and loss. Not healthy feelings to have lying around in a relationship.

The framing of sex as something sinful limited medical research as well as female pleasure. In 1620, John Moir summed up the problem of being a student of anatomy. "A consideration of the genital members is very difficult, and everything should not be revealed particularly with youths, because sin makes the subject of generation diabolical and full of shame, and a discussion might excite impure acts." This, then, would excuse some of our early anatomists for relying on previous wisdom, rather than investigating for themselves.

In his 2016 publication *Amoris Laetitia* ("The Joy of Love"), Pope Francis wrote of conjugal sex that we should not consider the erotic dimension of love simply as a permissible evil or a burden to be tolerated for the good of the family, but as a gift from God that enriches the relationship of the spouses.[122] He acknowledged that procreation is not the only reason couples have sex—take for example older couples, where the woman is post-menopausal. Natural family planning is advocated as a form of contraception, and he placed emphasis on the importance of mutuality in sexual relations. Maybe the future of sex within the Catholic framework *is* softening slightly?

Not only has the Church in former centuries limited intercourse to a procreative agenda but it has dictated how you should be practicing it, and this potentially affected the clitoris. Across Europe from the sixth through the twelfth centuries, medieval priests who heard confessions relied on manuscripts called penitentials, in which all manner of sins were listed, along with their appropriate penances. Many copies show signs of heavy use. Cambridge's Corpus Christi College holds the *Canons of Theodore,* a penitential handbook that lists oral sex as "the worst evil," requiring repentance by those who have indulged in it "up to the end of their lives." On the topic of masturbation an appropriate penance for a woman who has confessed to indulging in self-pleasure would be repentance "for one year." Although "one penance applies to a widow and a virgin; more is earned by her who has a husband if she fornicates." Woe betide the woman who forsakes the penis available to her in favor of her own hand. Penance for the masturbating man? Three weeks.[123]

A good friend of mine told me about having to go to confession at her Catholic high school in the 1990s, where she dutifully confessed to abusing her body—the coy term used by the Church for masturbation. She was given one Our Father and multiple Hail Marys for her sin. "Everyone was asking me what I'd done to deserve it," she said. "But I was too embarrassed to say. Years later when I told a boyfriend, he said, 'What? Why did you confess to that! No one confesses to it.' How did I miss that message?"

It became an increasingly natural expression of the hierarchical order of the world for men to take the dominant top position in intercourse, with the woman lying beneath. Although penitentials fell out of fashion, by the Renaissance, Tomás Sánchez, a Spanish Jesuit famous for his deductive moral reasoning, was declaring in his *Disputationum de Sancto Matrimonii Sacramento* that riding cowgirl was a mortal sin—although these were not his exact words. Of course not everyone would have followed the rules exactly, as books like *Aristotle's Masterpiece* show, but in religious circles, the clitoris might as well have not existed.

The Catholic Church has not been the only driving force in women's moral and sexual lives. In 16th-century Reformation Europe, Protestant priests like Martin Luther held that sex was designed to be a pleasurable experience between a man and a woman in matrimony.[124] Luther also said there was no evidence that the Bible prescribed a vow of celibacy for the clergy and it was his view that allowing priests to marry saved a whole load of sin, shame, and scandal. He was probably right, although the Church always taught that moderation was the godliest way to live. Unfortunately for a woman under Protestantism, while she was now free to engage in sexual intercourse within marriage—guiltless and, if she wished and could work out how, sans baby—the narratives of the moral nature of women and how sex and sexual pleasure worked had already created a hostile framework.

The idea that sex isn't acceptable unless it is practiced in a marriage committed to procreation has held strongly. The debate and strictures around prescription of the contraceptive pill when it was introduced in the U.K. are testament to this. When the pill was first made available to women in 1961, it was stipulated that candidates should be married and have had children. There was concern that prescribing it to younger, married but childless women would encourage sex and not procreation. This was not driven by religious beliefs, but by an agenda among decision makers—in this case male politicians and doctors—that sex should be firmly situated within marriage because marriage was good for social stability, just as procreation was good for the economy. The world needed workers and carers, just as the armies of Sparta and Athens, millennia earlier, had needed soldiers, and—in the eras of expansionism—colonies needed colonizers. The pill, it was perceived, would cater to the "needs" of men who'd done their duty by having children; their wives had perhaps been advised by doctors not to have any more for health reasons. These male needs could reasonably be met by prescribing the pill to wives to prevent dangerous pregnancies.

This stipulation was lifted in 1967 and the contraceptive pill became available to all women. In the space of six years, the debate about the role of intercourse shifted significantly. In the United States the pill was licensed for prescription to married couples in 1965, but it wasn't until 1972 that it became legalized as birth control for all women, irrespective of marital status. I'm struck by the fact that until recently men's sexual "needs" were seen as worthy of recognition and response; in contrast was the unspoken implication that women did not have sexual needs, and if they did, they were not important enough to be provided for.

The insistence by the primary religious force in European and American history that sex was exclusively for procreation, its passionate anti-masturbation message, and the backdrop of erroneous science all contributed to the marginalization of the clitoris in women's sexual lives—and I am concerned that it's happening more than ever before. The rigid, heteronormative insistence that P-in-V sex is the defining sex act masks the reality of lived sexual experiences. Today's teens have had more oral and anal sex than any other generation. And teenage girls are giving more than they are receiving. Peggy Orenstein reports that oral sex has become third base, and anal sex the consolation for preserving the chastity of the vagina. We need to be having conversations with these sexually active teens about safety, pleasure, consent, mutuality, and emotional well-being, no matter what type of sex they are having.

Is it all Eve's fault? What else was happening in the Western world to stigmatize the moral and sexual capacity of women?

Greeks, Romans, and Brothels
What have they got to do with the clitoris?

Greek writers and philosophers are a source of some profoundly damaging writings about the nature of women. Educators from the Renaissance to the 20th century assumed that the well educated would all have read the same books and shared the same grounding and language as most of the Western world's formative thinkers. It was believed that with a good knowledge of the ancient writers (a discipline sometimes known as classics or greats) you could become a leader of empire, a captain of industry, a general, or an admiral. You could do anything as long as you had knowledge of classical antiquity and a good, correct Latin style behind you. Indeed, it was only in 1960 that Oxford and Cambridge ceased to require a pass in Ordinary Level Latin as an entry qualification for undergraduate students. If you imbibed the greats, maybe you learned to think like them—and if that was the case, you'd believe women were substandard citizens.

The eighth-century BCE writings of Hesiod retell stories about the origin of the world, and one of the myths explaining the existence of evil relates to Pandora. Pandora was made by Zeus to be irresistibly attractive but, like a flawed avatar, she had "contrived within her lies and crafty words and a deceitful nature."[125] The beguiled Epimetheus married her. As a wedding present, Zeus gave Pandora a jar full of gifts, but told her she should never open it. Guess what? She couldn't resist temptation, and by her female hand all the evils of the world came tumbling out. The difference between Pandora and Eve is that Pandora carries no sexual guilt with her, although they both share the crimes of intellectual curiosity (greatly valued by academic institutions and employers these days, as mentioned before) and not obeying poorly explained instructions from authoritarian men. Most modern parenting books would advocate clear explanations to help children understand consequences, but neither Eve nor Pandora were given this benefit.

Aristotle was very clear about women's limitations:

> Wherefore women are more compassionate and more readily made to weep, more jealous and querulous, fonder of railing, and more contentious. The female also is more subject to depression of spirits and despair than the male. She is also more shameless and false, more readily deceived, and more mindful of injury, more watchful, more idle, and on the whole less excitable than the male.[126]

The role of brothels

There was a definite gender split by classical times in terms of what society considered acceptable sexual behavior by men versus what was acceptable in women, and women's lives became constrained by this. In Greek and Roman cultures, male promiscuity was considered normal and healthy, but female chastity was required for respectable women to ensure the integrity of family bloodlines. Both civilizations needed soldiers and leaders, so women were increasingly used as breeders. In Rome, Emperor Augustus (63 BCE–14 CE) took marriage-bed duties very seriously, passing laws insisting that all men between 25 and 60 years of age and all women between 20 and 50 were to marry and have children, or pay extra tax. Men were urged to have intercourse with their wives three times a month, but they were free to visit prostitutes and to take other lovers, in whatever form they came. The fact that marital intercourse was enshrined in law suggests that it was a less-than-popular activity. Child-rearing is a serious business in the creation and sustaining of an empire. (Queen Victoria, monarch of the largest empire to exist, was a much later historical model of pronatal womanhood. She had nine children.)

Chaste daughters became valuable to their families in the matrimonial trading game. Poor or slave women found their sexual freedoms restricted in another way, by the creation of brothels. In sixth-century BCE an Athenian lawyer called Solon set up the first brothel. Sex became a male-owned commodity with a role in the political and economic system. Brothels were good for raising taxes and they kept valuable wives, mothers, and daughters safe from unmarried soldiers, sailors, and traders. Arguably there were three long-term consequences for the clitoris.

Female orgasm didn't matter: The requirement for a man to pleasure a woman in a brothel was diminished by the fact that he had paid for the sex. Maybe men ceased to prioritize women's sexual needs over their own, and women didn't have the forum to ask for more. This attitude toward women and sex might have undermined married life, where wives already found it expedient not to be too enthusiastic about sex, as it raised questions about their suitability as good wives.

A similar lessening of regard for female sexual pleasure is shown centuries later in the popular 1960s and 70s book *Love and Orgasm*[127] by Alexander Lowen, who said he did not like to recommend clitoral stimulation to his patients because to most men "the need to bring a woman to climax through clitoral stimulation is a burden." Lowen sympathized with men on this matter and said providing clitoral stimulation before their own orgasm "imposes a restraint" upon their "natural desire for closeness and intimacy." Providing a woman's orgasm postcoitally, he said, would interfere with the "relaxation and peace that are the rewards of sexuality." Lowen concluded by saying, "Most men to whom I have spoken who engaged in this practice resented it." Mr. Lowen, I beg to differ. I believe that most men genuinely like their partners to experience orgasm. I'm prepared to wager that many heterosexual men find a woman orgasming, or the idea of it, a thorough turn-on.

It pays to fake it: Women obliged to provide sex in brothels probably found their services rated more highly if they were seen to "enjoy" the process. They might have found it expedient to flatter their clients and feign orgasms. Women who fake orgasms give two reasons for doing so: either the desire to hasten the end of the sexual encounter, or the wish not to make their partner feel inadequate. Both would be valid reasons for a sex worker to fake it. An outcome might be the belief that female enjoyment of sex was part and parcel of women living sexually promiscuous and morally questionable lifestyles, and this type of pleasure would be something undesirable in wives.

In surveys, most men say that during their sexual encounters, the women they're with experience orgasm. But the 2015 *Cosmopolitan* sex survey found that more than two-thirds of female respondents admitted to having faked orgasms with partners, and other surveys show similar results. The numbers don't add up, folks. Someone is not being honest. And these men are either: a) lying, b) deluded, or c) telling the truth. It's like the research that shows the majority of people consider themselves to be better-than-average drivers compared to other road users. We cannot *all* be better than average. So, guys, let me suggest that if you need to ask the question "Was it good for you?" the answer is probably no—but at least she didn't fake it. Rather than ask whether it was good, why not ask, "How could it have been better for you?" You are doing the most intimate thing together, and while starting a conversation can be hard, it has to be worth it. Recent research found that women who are comfortable saying the word *clitoris* are also the women more likely to be experiencing orgasm in their heterosexual encounters—so get comfortable with saying it, ladies, so you can ask for what you want. And guys, if you are engaging in an intimate activity and your partner moves ... notice the move and stay with it! I once saw a comedian whose skit was about a disastrous sexual encounter, where every time she adjusted herself so that the stimulation she was receiving was more effective for her, the guy readjusted himself back. From the knowing laughter among the women in the audience, you could tell she'd struck a nerve, unlike her lover.

In 1884, in his book *Hygiène et Physiologie du Mariage*,[128] Auguste Debay advised women to fake orgasm because "man likes to have his happiness shared." As recently as 1995 a writer for *Cosmopolitan* endorsed this behavior as "just a matter of expediency, not to mention common courtesy."[129] Ladies ... be honest! You are not doing anyone a favor by faking it. Guys, please just ask early on in a relationship, with no judgement on her or on yourself, about what works for your partner. Who wants their orgasm to be a faked courtesy? Shouldn't it be a central part of any mutually satisfying sexual encounter?

There are two types of women—wives and whores: Women could now be categorized as either chaste, wifely material or sexy whores. Roman law helped reinforce this dichotomy. If you committed adultery with another man's wife or his chaste daughter, you were on the wrong side of the law. Sex with slaves, prostitutes, or those on the margins of society (criminals, entertainers, gladiators, for example) was fair game. When Christianity was assimilated into Roman culture, the New Testament played to this polarization nicely, offering up the Virgin Mary and a caste of loyal mothers versus Eve and the Whore of Babylon, "mother of harlots and abominations of the earth," who is portrayed as active in her role as harlot, rather than a passive victim of subjugation. With her "the kings of the earth have committed fornication, and the inhabitants of the earth have been made drunk with the wine of her fornication."[130] The Whore of Babylon's fornication (intercourse outside marriage, therefore against the laws of God) is like wine. The metaphor implies that she makes people drunk with her wine, so that they lose their judgement and control. The drinkers are passive, and this is their get-out-of-jail-free card. "She made me do it." "She led me on." It's been a convenient ruse for men who evade guilt by indulging in this convenient blame-shifting. It's always easier to lay the blame for guilt and feelings of shame at someone else's cunt. Sorry, I meant to say *door.*

Early Humans
Did the clitoris ever matter?

We don't know. Archaeologists excavating and exploring prehistoric sites have found representations of the vulva.[131] Some Paleolithic versions make a feature of fissures in the stone, others are carved into rock faces, and some you can hold in the palm of your hand. The vulva is the focus, whether it is disembodied, or part of a female form as in the second image.

Two of three identical vulvas, thought to be 32,000 years old, found on the wall of the Abri Cellier rock shelter that runs along the Vézère Valley in southwestern France, can be seen in this image.

Vulva carvings from Abri Cellier, France.

A cave at Bedeilhac in the French Pyrenees, famous for its stalactites, contains a life-size vulva finger-drawn in clay that even has a small stalactite clitoris. (www.donsmaps.com is an easy resource.) Search on *the vulva in stone age art,* and you will be surprised at how many similar references come up.

If you think these carvings are ambiguous, try looking at the Venus of Monpazier (also from the Dordogne, circa 25,000 BCE), which leaves no room for doubt. There is nothing abstract about her vulva. It is her defining feature.

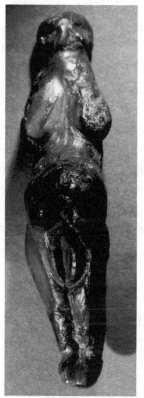

Venus of Monpazier, Dordogne, France.

Imagine you are an archaeologist. Which of the following would you choose as a theory to explain these Stone Age vulvas?

a) They are examples of Palaeolithic pornography.
b) They were for sex-education purposes.
c) They record or celebrate the life-giving capacity of women.
d) They are an anomaly, and the male versions haven't all been found yet.
e) They are examples of graffiti.

Until recently the most popular theory was a), pornography . . . because what else could they be? We all know that a pussy is pornographic, right? Why else would cave*men* have made them? Actually, academics are now suggesting these carvings celebrate the power of the vulva as a life force, which makes sense when you think of all the creation myths that start with a woman. We don't know the purpose of these representations or the gender of those who made them. Maybe it was teenage-boy graffiti, as suggested by R. Dale Guthrie in his book *The Nature of Paleolithic Art*[132] (although in my experience teenage boys are more likely to draw erect penises than vulvas). Maybe it was cave*women* rather than cavemen who carved them? They may have been for quasi-religious purposes, or they may simply be a record of the most significant things in primitive societies, such as fertility. However, we do know that for many eons across Europe the vulva was important enough that as soon as humans could make representations, it was on their list of things to represent. And, once in a blue moon, a clitoris was included or hinted at. Which is more than you can say for art today.

The transition from the Paleolithic to the Neolithic era around 10,000 BCE undoubtedly saw significant cultural shifts for humans in the ways they led their lives. Some groups of humans gave up the nomadic hunter-gatherer lifestyle and began farming. These new societies had more people living together than previously, and with this came the need for rules to regulate human conduct in a different way. There was increased personal wealth (a home, land, livestock, tools) to be guarded and inherited. In their book, *A Women's History of Sex,* Harriet Gilbert and Christine Roche[133] argue:

> The more that a man is concerned with his property the stricter will be the rules by which he encircles "his" women's sexuality. While the dispossessed poor have usually condoned both pre- and extramarital sex—for women as well as men—daughters of the rich have been married off as children to ensure that they were still virgins, and, once married, kept locked in the house, the castle, a chastity belt.

We don't know when a culture of restricting women's sexuality started, but it seems tied to the moment when there is wealth to protect. Nor do we know whether a culture of sexual jealousy or control existed prior to this because we don't have a record of the lives of women in prehistoric times. Even if we did, it would be like future generations finding only *Love Island* or *The Bachelor* and making assumptions about sexual behavior and pair bonding in 21st-century society. Archaeological research can examine human remains and settlements to establish what people were eating, what kinds of injuries they sustained, what kind of repetitive movements their lives generated; but archaeologists can't tell us about their sex lives or attitudes about the clitoris. For the purposes of this book, the best we can do is to stick to what we do know. By the first millennium BCE, ancient Hebrews were running with the theory that men were more significant in making babies than women. Women were reduced to carriers. According to Philo of Alexandria:

> The matter of the female in the remains of the menstrual fluids produces the fetus. But the male [provides] the skill and the cause ... the male provides the greater and more necessary [part] in the process of generation.[134]

The idea that women were merely baby hosts remained "fact" until the 17th century and continued to be the dominant theory until the 19th century. The discovery around 2,000 years ago that men were the originators of human life, with women taking an arduous supporting role as facilitators, positioned men as the hero and the penis as the genius for a *looong* time.

As an aside to this, when we announced that we were having our first child, most male friends responded to my husband with, "Well done, mate! Not firing blanks, then!" without a word of acknowledgment for my role in the process. They were often affronted when I retaliated with, "I did some of the work too, you know. I provided the egg, and I'm the one growing the baby." Chapter 5 of the book of Genesis becomes a catalog of fathers begetting sons, for 35 verses, stretching out patrilineal descent from the dawn of time onward, with not a mention of the mothers. It may be that at this point in prehistory, womanhood as a symbol of fertility was downgraded, and women took up their new role as baby hosts. This could have been a determinant in women becoming lesser stakeholders in society and in their own sexuality. The most significant factor for the clitoris in this history is her absence.

Ten thousand years ago, the early years of human settlement ushered in a prioritizing of all things male in terms of societal organization. Gilbert and Roche argue that while pre-Christian times limited the freedom of women and put rules around sex, they didn't deny sexual pleasure. They quote the Old Testament Song of Songs: "I am my lover's and he desires me." And Ovid's lovely advice in *Ars Amatoria* is, "When you have found the place where a woman loves to be fondled, don't be ashamed to touch it any more than she is ... be sure that you don't sail too close too fast and leave your mistress behind." Sound guidance for a lover. But there was such thorough disregard for women and high esteem for female chastity that by the time Christianity was made the official religion in the Western world, in 323 CE, women were disempowered and there was little they could do to protest against the denial of their sexual pleasure. The clitoris, already being eviscerated by dominant anatomists, was now also ghosted by religion. The Church built on broken ground.

Philosophers like Aristotle set the tone for women to be perceived as immoral and weak willed, and for centuries men's theorizing about the nature of women and the ideal woman did not take a woman's perspective into consideration.

Rousseau, he of "Man is born free and everywhere he is in chains," and who inspired the French revolutionary cry "Liberty, equality, fraternity," is a prime example of how concepts about women and their sexuality became embedded in European and American cultural thought. Fraternity was not a euphemism for siblinghood but was gendered, and the sisterhood was excluded. Rousseau preached that the role of a woman was to "be unknown"—she should take reflected glory only, from her husband or children. Woman, she was to stay in chains. He also urged that the ideal woman should be "natural." Rousseau was responding to the decadence of France's *ancien régime* and Europe's high society at the time, but the feminine naturalness he advocated was also an artifice. Rousseau defined this trait as being sexually passive, having a desire for pregnancy and breastfeeding, and a dislike of reading, writing, and learning; he favored a love of housekeeping, sewing, gentleness, selflessness, and arranging oneself to be attractive to men, with a demure display of the curve of one's breast (that's an oxymoron if ever there was one). It is a reflection of what *he* wanted to be natural. It's a slippery adjective.

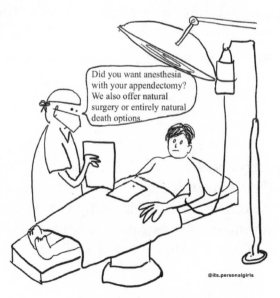

I went to a highbrow source to understand Rousseau's influence. The *Stanford Encyclopaedia of Philosophy* has this to say:

> Jean-Jacques Rousseau remains an important figure in the history of philosophy, both because of his contributions to political philosophy and moral psychology and because of his influence on later thinkers.[135]

Rousseau perpetuated a vision of the ideal woman as passionless and selfless. I don't know about you, but here are *my* responses to that:

a) Women are not passionless.

b) Women don't have to be selfless. It's O.K. to put oneself on the agenda. This is not something many women have felt free to do in sexual encounters, nor is it something women have been good at in myriad walks of life, because society has not encouraged them to develop a strong sense of self. Self is *not* a binary choice between selfless and selfish; it is both of these and every nuance in between, and the balance changes constantly in response to personal needs combined with external factors.

c) Deviation from Rousseau's vision left women open to accusations of being *unnatural*, a word used to denote immorality and forces of evil, like the witches in *Macbeth*.

This part of the book started with a quote from *Catch-22* that illustrates the impossibility of the central character's situation. Orr is keen to avoid war by being sectioned out of the army. He can only be sectioned out if he is insane, but if he were insane he wouldn't have the wit to seek such medical help. So he can't win. Women and their sexuality have for centuries been caught in a similar catch-22. Enjoy sex too much, ask for better sex, give yourself an orgasm, and you would be considered immoral—which does nothing for your chances in life, or for the advancement of women's equality.

Fear stems from misunderstanding, and there has been profound misunderstanding of women's sexual anatomy and a deeply ingrained misunderstanding of women as equals—but we no longer operate in a world where this is the case. Anatomical knowledge has never been so advanced, and equality may be resisted but it is not up for debate in informed circles and has many champions. We (the siblinghood, not just the sisterhood) need to speak up when we hear misinformation, check those who use language to make moral inferences about women and their sexuality, and get used to putting a woman's orgasm into the heterosexual sexual experience.

Throughout the centuries, scientists, religious leaders, and philosophers have all held strong views on the nature of female sexuality. They have sought to understand the anatomical makeup of woman; the right way to live life as a woman in relation to God; and the essence of woman by a philosophical process of deduction. They have deemed themselves qualified to speak about their findings using fact, opinion stated as fact, illogic, and a lot of rhetoric to become authoritative and influential in defining the social systems within which women function. The early 19th century heralded the emergence of a new field of academic inquiry that would be used to further scrutinize women and their sexuality. Enter the new science of psychology.

4
Psychology and Sexology
When do women get a say?

"There is nothing either good or bad, but thinking makes it so."

Shakespeare, *Hamlet*

Since Freud, psychologists have been telling the world how the female mind works and how it affects female sexual behavior. What's wrong with this? After all, a lot of sex is in the head—for example, attraction, anticipation, getting and being in the mood, pre-sex arousal, and fantasy sex. Some fascinating contemporary research, involving brain imaging combined with reported and monitored arousal states, is beginning to tell us a lot about the similarities and differences among individuals when it comes to sex. It seems Freud was right about some things. But he also messed up: a lot of what he said about female sexual arousal and the clitoris was plain wrong, his scientific theory on those subjects was not grounded in scientific methodology, and he presented this ridiculous theory as a proven fact.

Freud and the immature clitoris

Freud's theory that subconscious drives or needs are often in conflict with our rational minds is well established and hard to disagree with. How many of us have made rational, well-intentioned resolutions in the morning, only to find we've overridden them by mid-afternoon? His theories on attachment, and the idea that early-childhood relationships are a major factor in the emotional development of children and adults, have been endorsed by subsequent psychologists. He knew a thing or two about human nature. His belief that a healthy relationship with our basic bodily functions in childhood is essential to our well-being formed the foundation of his well-known theory of personality, and this is why his thoughts about your clitoris matter—because he introduced the idea that it could account for the development of your personality.

There are five stages of development in the transition from baby to adult, Freud suggested, and successful progress through these stages is necessary to become a balanced, untroubled adult—not just in the ability to function as sexual beings, but as complete emotional and rational citizens. These recognizable stages started, he said, with the *oral* stage, which anyone who has spent time around a six-month-old can relate to. Hand–mouth coordination is achieved, teething is starting, and everything the child can grasp is thrust into its mouth: its toes, its toys, its food, the dog's food, the dog's toy, dirt, sand, you name it. This lasts until around 18 months old, when the child transitions into the second, *anal* stage, which is primarily related to developing healthy toilet-training habits. Between three and five years old, Freud said, a child goes through the *phallic* stage, when children need to develop healthy substitutes for the sexual attraction boys and girls have toward a parent of the opposite gender. This is when the Oedipus or Electra complex might raise its head and is where I begin to fall out with Freud. Love for a parent is not sexual love; why would the two necessarily be conflated? Stage four is the *latency* stage, which Freud thought initiated the development of healthy dormant sexual feelings for the opposite sex. Finally, from the age of 12 onward, the growing child enters the *genital* stage, in which all tasks from the previous four stages are integrated into the mind, allowing for the onset of healthy sexual feelings and behaviors. Freud argued that a major factor in successful progression through these childhood phases was resolving the conflicts between physical drives and social expectations that arose during these stages; if this didn't happen for any reason, he said, then problems would appear in adult life.

What are the implications for women and our clitorises? Firstly, during the phallic stage, Freud suggested, girls experience penis envy when they discover that women's and men's anatomies are different—and this awareness causes them anxiety and envy as they realize they were either shortchanged by evolution or have been castrated. It's a lot for a three-to-five-year-old girl to take on board. It's a miracle that girls are well adjusted at all. Let's unpack these penis-envy theories in more detail.

An inferior penis: Girls and women were supposedly shortchanged by evolution because they only got a tiny penis. Freud said, "As regards little girls, we can say of them that they feel greatly at a disadvantage owing to their lack of a big, visible penis, that they envy boys for possessing one."[136] Freud, like some anatomists before him, maintained that the clitoris was "homologous to the penis" but he was also clear that it was "inferior" to a penis, because the clitoris was never going to grow. Returning to my introductory bath-time story, my sons' verdict on my daughter's discovery of her clitoris was along Freudian lines. They found her statement risible. "That's not a penis! It's not big enough," they insisted. Ever an advocate for women, my three-year-old daughter was equally adamant and shouted back, "That doesn't matter! It is!"

An inferior superego: Apart from giving the clitoris an inferiority complex, the damaging aspect of Freud's belief is that he tied this clitoral inferiority to his theory of the development of the superego. The superego is the part of one's three-part psyche that incorporates the values and morals of society, and it starts developing in childhood. A fully developed superego is what allows one to be a functioning, capable citizen of the world; it is, if you like, a moral conscience. The id is the aspect of the mind that contains urges or drives, like sex or aggression, and hidden memories. The ego is the realistic part that mediates between the desires of the id and the superego, where societal and parental guidance is weighed up against personal pleasure and gain. Freud suggested that a woman's lack of a penis, and her discovery of this in the phallic stage, affect the development of her superego[137]—not her ego, which would be her ability to mediate between her desires and what is socially right, but her *fundamental capacity* to understand a social moral framework.[138] This, he reasoned, accounts for woman's moral inferiority. Here we go again: the moral inferiority of women is a given, and all psychology had to do, like science and philosophy before it, was explain why. We should be furious. When Mark Twain wrote, "The very ink with which all history is written is merely fluid prejudice," he wasn't wrong.

A missing penis: The castration theory forms the second part of Freud's hypothesis about the phallic stage in young girls. On discovering biological sex differences (mainly the issue of their missing penis), girls believe they once had one but someone, probably their mother, cut it off. Yikes! Where did that come from? Really? I know my daughter was looking for her penis, but her clitoris did her just fine once she found it.

A girl's development, Freud argued, is forever after hindered due to her lack of a penis. Because of her stunted superego she is not equipped for public and political existence. This a girl can never make up for. Her best hope for a second-rate maturity is to shift her focus from her clitoris to her vagina. He claimed that "[t]he elimination of clitoral sexuality is a necessary precondition for the development of femininity since it is immature and masculine in nature."[139]

Freud acknowledged that the clitoris was a site of sexual pleasure, but then insisted that it be sidelined in favor of the vagina, rather than allowing the two to exist in pleasurable harmony. He not only switched the clitoris out, but stigmatized it, and made it the reason women were destined to be second-class citizens.

In his 1916–17 *Introductory Lectures on Psycho-Analysis*, Freud said:

> In her childhood, moreover, a girl's clitoris takes on the role of a penis entirely: it is characterized by special excitability and is the area in which autoerotic satisfaction is obtained. The process of a girl's becoming a woman depends very much on the clitoris passing on this sensitivity to the vaginal orifice in good time and completely. In cases of what is known as sexual anaesthesia in women the clitoris has obstinately retained its sensitivity.[140]

Freud insisted that adult clitoral sensitivity led to sexual dysfunction. Any forward-thinking men or women of the time, keen to engage with contemporary thought, were sent back down the route of the vagina as the focal point of mature female sexuality, leaving the clitoris behind as a sad vestige of a penis, in the realms of childish pleasures, or as a symptom of sexual dysfunction. Any parent who wanted their daughter to develop appropriately would be keen to wean her off her clitoris. And a woman who failed to "achieve" orgasm through coitus was labeled frigid. This all tied in very neatly with the scientific and cultural narratives that already existed about how sex worked, and the moral inferiority of women. Furthermore, Freud reasoned, a mature woman could legitimately have a penis, albeit temporarily, through sex, and in the longer term through the conception and delivery of a boy—or as he called it "a penis baby." I wonder how my sons are going to feel when I next greet them with, "Hello, my lovely penis babies!"

While acknowledging that the clitoris was a site of "special excitability" and "autoerotic satisfaction" for children, Freud failed to acknowledge it as a viable source of sexual pleasure in womanhood. He believed that, for women, sex was not at heart about pleasure. Women were, in his words, "passive receivers." Freud understood many things about the human psyche, but he did not understand women. Instead he demonstrated a sound understanding of centuries-old prejudice and misinformation.

Contemporaries—notably Karen Horney, a fellow psycho-analyst—criticized Freud for his theories about penis envy, but he headed off this discussion, apparently writing in a much-quoted essay, "We shall not be very greatly surprised if a woman analyst, who has not been sufficiently convinced of the intensity of her own wish for a penis, also fails to attach proper importance to that factor in her patients."[141] His dismissal is as damning as Vesalius's of Colombo's discovery of the clit.

To some extent, I understand why women might have had penis envy in Freud's day. In an era when women were petitioning for suffrage, the right to be heard was worthy of envy. As a metaphor, the concept of penis envy works; as an explanation for the way women work sexually, it does not. You can see why this moment in history, less than a hundred years ago, was a low point in the clitoris's history. She was blamed for making women less morally able than men, and as a site of pleasure she was damned as immature—just as anatomical research was getting to the truth.

In 1933, toward the end of his esteemed and influential career, Freud must have been shaking his head as he wrote, notoriously:

> That is all I have to say to you about femininity. It is certainly incomplete and fragmentary and does not always sound friendly . . . If you want to know more about femininity, enquire of your own experiences of life, or turn to poets, or wait until science can give you deeper and more coherent information . . . Psychology cannot solve the riddle of femininity. The solution must, I think, come from somewhere else . . . We know nothing whatever about the matter.[142]

What?! Now you tell us? Frankly, Freud, if you didn't know what you were talking about, why TF did you start talking in the first place? Shame on you.

@ its.personalgirls

How did Freud's theory impact female sexuality?

Freud's doctrine reinforced fear about the clitoris and allowed people to continue to insist that the vagina, with the essential help of a penis, was the right and proper place for an orgasm. It also gave the Western world another reason to believe that women were not well equipped for public life. During the Freud years, many women spent hours in therapy trying to shift their focus from their clitorises to their vaginas, and some underwent surgery to cure them of their "immaturity." According to Helen Singer Kaplan—who founded the first medical-school-based U.S. clinic for the treatment of sexual disorders, and was a leader among scientific-orientated sex therapists—until the practical study by Masters and Johnson in 1966, "most clinicians believed that stimulation of the clitoris produced 'clitoral' orgasm only in infantile women, i.e. those who were fixated at an early stage of development and had failed to achieve genital primacy. In short, retention of clitoral sensation was considered prima facie evidence of neurosis."[143]

Freud's legacy lived in the sexual consciousness for much of the 20th century. The First and Second World Wars interrupted sex research and reset agendas, with the next big foray into human sexuality being led by Alfred Kinsey in the 1950s, although, as Kaplan's comment above highlights, his discoveries (about female sexuality, at least) went largely untapped.

Kinsey lets the genie out of the bottle.

Alfred Kinsey was a zoologist who studied human sexuality in the same clinical way that he had previously studied gall wasps. He was interested in behavior rather than motivation. He gathered information and reported the data as he found it, contextualizing it within the world of animal behavior. He was most interested in sexual "outlet" (orgasm), and this is what he centered his data collection on. *Sexual Behavior in the Human Female,* published in 1953, was based on the case histories of

5,940 women. Kinsey's conclusions were that women masturbate, women fantasize about sex, women are not much slower
to reach orgasm than men when masturbating, vaginal walls
have few nerve endings, masturbation focuses on the clitoris
and labia more than on penetration, a substantial number of
married women did not experience orgasms, and pre- and
extramarital petting and sex was taking place. Also, those
women who participated in premarital petting were more likely
to have marital orgasms. The genie was out of the bottle.

In 1954 Kinsey's annual funding from the Rockefeller
Foundation was canceled due to pressure from religious groups.
Some reviewers were damning. In a review titled "Behaviorism
with a Vengeance," Dr. Abram Kardiner, a clinical professor of
psychiatry, described the data as "prurient." He was outraged
that "with very little trouble any woman can thumb through
this book and come out with a chart of her own personal
standing in the great norm that is herewith established."[144] Dr.
Kardiner criticized Kinsey for underreporting female frigidity,
and complained that "by making the clitoral orgasm the basis
for his standard and refusing to acknowledge the difference
between this and the vaginal orgasm, many a woman in his
sample is recorded orgasmic on a technicality." A *technicality!*
Kardiner ended his review by dragging women back to the
sex-moral imperative, saying sexual morality is "one of the main
supports of the entire structure and functioning of society."
He argues that Kinsey's "misinformation is dangerous and can
only have one outcome. It will add still more confusion to the
sexual unrest of our time and for this, not Dr. Kinsey, but the
public will pay the price." The morally outraged were intent on
putting the genie back in the bottle.

Masters and Johnson discover there are multiple genies.
In the late 1950s, William Masters and Virginia Johnson liberated another genie. They pioneered sexual research by directly

observing the anatomical and physiological sexual responses of about 700 copulating couples and masturbating subjects. Masters and Johnson noted that women's orgasms were physiologically the same as men's orgasms. They attributed orgasm in women to the clitoris and identified four phases of the human sexual-response cycle (excitement, plateau, orgasmic, and resolution) giving women sexual parity. What's more, they found that women were capable of multiple orgasms, without the refractory period that men needed. Now there were multiple genies out of the bottle.

Neither Kinsey's nor Masters and Johnson's research is without problems. One recurring issue is how to create a random sample, because in any truly random sample there will be subjects who have no desire to reveal the intimate details of their sexual experiences. Unless your sample is truly random, critics argue, the sex-crazed, deviant, and those prone to exhibitionism might be the only ones to respond. Also, you should randomize by race, region, age, and socioeconomic and educational factors otherwise your research is deeply biased. The second issue undermining sex research is the innate biases of experimenters themselves. Kinsey had a behaviorist bias— he saw behavior as a function of animal drives sometimes repressed by society, and was progressive in his attitudes to all thing sexual; Masters and Johnson favored heterosexual sex as the ideal forum, betraying their deep cultural conservatism. They selected for their studies couples where the woman was orgasmic in intercourse, and while they concluded it was because of the clitoris and its stimulation during intercourse, they failed to recognize this was not the case for most women. By tapping into these flaws, detractors were able to undermine the validity and question the robustness of the research and its findings about female sexuality and the clitoris.

Hite: The genie gets an ally, but is she the Fairy Godmother or Maleficent?

It was no different for Shere Hite, who in 1976 published *The Hite Report: A Nationwide Study on Female Sexuality.* Hite distributed 100,000 questionnaires across America through national mailings to women's groups, notices in magazines like *The Village Voice, Mademoiselle,* and *Brides,* and church newsletters. Her questionnaire asked for intimate and specific details of women's lived sexual experiences. I love the last question, no. 58: "Why did you answer this questionnaire? How did you like it?" It expresses a real interest in the 3,000 women who took the time to write detailed answers to the previous 57 questions—and their responses demonstrate their desire to speak up and be heard. It was the first time a researcher had asked women in complete anonymity how they felt about sex, what they liked about sex, and how they behaved in sexual encounters. Kinsey interviewed his respondents; Masters and Johnson observed theirs. Nine-tenths of Hite's 400-page book consists of verbatim quotes from the respondents, topped and tailed with summarizing and concluding commentary by Hite. This must be some of the rawest data from female witnesses of their sexual experience.

From the perspective of the clitoris, two aspects of this landmark report are of interest. Firstly, the material it generated. Only five of its fifty-eight questions specifically referenced the clitoris or clitoral stimulation,[145] yet the respondents constantly mentioned the clitoris—in answer to questions about how they best achieved orgasm, how they masturbated, and what they wanted more of in sexual encounters. Secondly, the strong reactions the report generated. People either loved it or hated it, and those that hated it did so with a vengeance, fixing on the unrepresentative sampling method to undermine its validity. Critics argued that only the lusty, man-hating,

masturbating feminist hussies of America rose up and replied.
Two from Little Rock, Arkansas; one from Boulder Creek,
California; one from Northbrook, Pennsylvania; one from Los
Alamos, New Mexico; one from Wolf Creek, Oregon. A smat-
tering of wicked women from every state sat down and wrote
subversive answers to the 58 questions about orgasm, sexual
activities, relationships, life stages, and feelings about these
experiences, all supposedly with the express intent of undermining the penis. What else could you expect from feminist
groups or liberal magazine-reading women? "Conspiracy!"
the critics cried. They pointed out that it was entirely possible
that the woman in Carbondale, Illinois, had colluded with
the woman in Claremont, New Hampshire, who had in turn
colluded with someone in Columbus, Ohio, and they with
the witch in Colombia, South Carolina, who had sucked in
respondents from Carol Springs, Florida; Carrollton, Georgia;
Cologne, Minnesota; and Cleveland, Texas. Apparently, these
women ganged up to deceive Miss Hite and led her to believe
that women were not enjoying coitus as much as their menfolk,
which, in the faultfinders' view, was a blatant lie. Obviously,
these detractors argued, the multitudes of good women who
passed over the opportunity to fill in the questionnaires did
so because they were orgasmically sated through intercourse,
couldn't see the purpose of the survey, and were properly
modest about this most private aspect of their lives. The thousands of unfilled questionnaires, they argued, spoke for themselves: these women had nothing to say about sex.

It would be fascinating to learn the reasons many women
didn't take up the opportunity to participate. Some we can
guess at: plenty of them certainly wouldn't have had time to
do it; some would have been uncomfortable discussing such
private and intimate details; some would have worried about
disapproval; maybe others couldn't see the relevance because

they were having amazing sex, or they had given up on sex, or they didn't think their perspective was important. It's true, the responses were not statistically robust, because they came from a self-selecting group of women who were prepared to engage with the topic of their sexual experiences and wanted to contribute to a wider discussion about female sexuality. They wanted to bear witness.

Having proved the quantitative failure of the project, critics damned the qualitative findings, and even suggested that Hite had invented some of the responses. An anonymous reviewer in *The Harvard Crimson,* the daily student newspaper from that university hotbed of learning, said, "The theories and quotations Hite presents fit together so smoothly, one cannot avoid suspecting her of manipulating information. I suppose I would have felt this less if some distinctly anti-male sentiments didn't frequently creep into the book." And the good people of Harvard were joined by *Playboy,* which used its voice to rebrand the book as *The Hate Report.* The report was not a piece of academic work to be taken seriously, they implied. Like women who cry rape. Our little clitoris was sent back into the shadows, not welcome in this new era of sexual liberation. It was all sheer hype.

I find the cohesion of the responses to Hite's questionnaires convincing: the details the respondents shared chime with women's experiences of sex as reported by contemporary magazines—and my friends—and the numbers, although not statistically robust, have been consistently echoed by subsequent surveys. Here is a top-line summary.

Orgasm

Of the sexually active respondents:

- 30% experienced orgasm regularly from intercourse.
- 22% rarely experienced orgasm with intercourse.
- 19% experienced orgasm during intercourse with simultaneous clitoral stimulation.
- 29% didn't orgasm during intercourse through any means.

Masturbation

Of all respondents:

- 82% of women said they physically enjoyed masturbating, although psychologically it left many feeling guilty.
- 95% of this group orgasmed easily and regularly with masturbation. Hite noted that "[m] any women used the term 'masturbation' simultaneously with orgasm: women assumed masturbation included orgasm."
- Only 1.7% masturbated through mimicking vaginal penetration.
- For the others, clitoral stimulation was a key feature of the activity.
- Some included vaginal entry as part of the experience.

Hite reported that 11.6 percent of her respondents never orgasmed, and most of this group also never masturbated. However, they did want to be orgasmic. These women tended to be more tolerant of non-orgasmic sex than their peers, who expressed frustration and feelings of anger and humiliation about sexual encounters that resulted in orgasms for their partners but not for them. Once you know what you are missing in a sexual encounter, it becomes an issue. My favorite quote from this section is the woman who replied, "Whoever said orgasm wasn't important for women was undoubtedly a man."

On the topic of what many in the past have called frigidity, while women wrote of learning to orgasm through masturbation—some discovering it early in life, many others well into their sexual lives—few had been able to translate its nirvana to a vaginal orgasm. Instead they reached orgasm with intercourse in other ways: furtively stimulating themselves during coitus, or masturbating after their partners rolled off and fell asleep. Some women wrote of shedding their inhibitions and bringing masturbation into their sexual repertoires with their partners, and for these women, the quest for a vaginal orgasm receded in importance. Some women didn't rate their long-term partners very highly as lovers, but loved them deeply and were resigned to poor intercourse experiences. Others were adamant that if their sexual partners weren't open to mutual pleasure, they moved on. However, many women owned up to faking orgasms during intercourse.

Of the women who answered the question "Do you ever fake orgasms?" 34 percent said yes. Another 19 percent said they used to fake them but then they thought, "Why should I?" Reasons women gave for faking were about boosting the morale of their partners ("to build his ego"; "he'd be disappointed if I didn't"; "to keep him interested") or to hasten the end of the interaction ("I'll never come anyway"; "to get it over with"). This has been echoed numerous times in subsequent research.

If you are curious about women and sex, *The Hite Report* is a relevant and humbling read. It would teach a man interested in women and their sexuality more than any porn or sex-education book can. It was an enormously important piece of research in the history of the clitoris. It told us what women did with it, what they liked others to do with it, and how they felt when the stimulation was cut short. It put the clitoris at the center of the female sexual experience in an era when women had been told they were sexually liberated. But it was not a report that brought about enlightenment. Most girls who reached puberty in the 1970s, who were not old enough to pick up on the media discourse about Hite, grew up to be none the wiser about her findings. The clitoris continued to be hushed up, referenced as an insignificant dot on anatomical diagrams, and not deemed worthy of discussion. Why did everyone believe Freud—but not Kinsey, or Masters and Johnson, or Hite?

Shere Hite's work was conducted 50 years ago and, apart from Dr. Ruth, there have still been few other big voices in the arena. John Bancroft, M.D., a former director of the Kinsey Institute, reported that sexual science lags behind other fields in terms of academic respectability and funding,[146] a view supported by researchers at the Society for the Scientific Study of Sexuality. Administrations that favor sex-education programs promoting abstinence do not create a fertile environment for such research. Naomi Wolf wrote an article for *USA Today* in 2012 called "Liberate Female Sexuality Research," in which she said,

> Forty years on, the data on female sexual satisfaction have not budged upward since Hite's day. In some ways they show erosion in pleasure and even a striking decline in many women's fundamental interest in sex ... Discourse about the value of women's sexuality and their erotic well-being has been so marginalized over the past few decades that in today's climate new

findings on female arousal and satisfaction are not being reported in mainstream media. When one brings them into public debate, as I have recently, I find that one must make the case from the start that these numbers—and female satisfaction—matter at all.

This is forcefully reflected by the findings of researchers Katherine Ellen Foley, Youyou Zhou, and Christopher Groskopf.[147] They downloaded 4,545 articles from *Journal of Sexual Research* and *Archives of Sexual Behavior*, the two journals that have served as the home for sexology research since 1965 and 1971 respectively and remain the most cited sources on the topic. Foley, Zhou, and Groskopf tracked the 1,000 most-used words in these journals from 1970 to 2017, and also tracked the frequency of use of these words over that time.

We found that over five decades, the most popular words in sexology evolved to reflect cultural ideas about what's normal bedroom behavior. Broadly these changes reflect major social events over time, including the sexual revolution, the AIDS epidemic, and the civil rights and LGBTQ movements. As sexual norms in the public eye evolved, so did the science studying [them].

They found that the words *bisexual, orientation, condom, gender,* and *risk* had increased in use over the five decades. *Masturbation, orgasm, coitus,* and *contraception* had all declined. *Penis* remained relatively constant in its use, declining by just 0.73. *Clitoris* didn't make the top 1,000 words.

5
Evolutionary Biology
What's the point of the clitoris?

"The challenge in providing an evolutionary explanation is to find a historical account of how a group of organisms came to have a particular property or trait."

Elisabeth A. Lloyd, *The Case of the Female Orgasm*

Clearly, sexual excitement is an inducement to have sex and an evolutionary advantage, but why did women's orgasms evolve? What's the point of the orgasmic capacity of the clitoris? I was hoping to uncover a Kipling-style story. My grandfather read the *Just So Stories* to me when I was a young girl, from a tatty, well-loved book he'd had since childhood. "How the Camel Got His Hump"; "How the Rhinoceros Got His Skin"; "How the Leopard Got His Spots." I imagined a "How the Woman Got Her Clitoris" narrative.

There is an obvious evolutionary explanation for a man's orgasm: mechanically it is essential to deliver sperm so that conception can occur, and the pleasure aspect ensures that procreative activity is sought after. Women's orgasms are less apparent in purpose. What are the most credible theories for the evolution of the clitoris?

Before evolutionary theory, anatomists and philosophers sought to understand how the body worked rather than asking why it worked in a particular way. As we've seen, early thinkers started from the first principle that women were inside-out versions of men, which made the clitoris hard to explain. However, while reading *Making Sex* by Thomas Laqueur I found the following explanations by Aristotle and Galen.

Aristotle built on the concept that body cavities need ventilation: the throat provides this service for the lungs and stomach, and the clitoris supplies ventilation to the vagina and womb, he said.

> The path along which the semen passes in women is of the following nature: they possess a tube—like the penis of the male, but inside the body—and they breathe through this by a small duct which is placed above the place through which women urinate. This is why, when they are eager to make love, this place is not in the same state as it was before they were excited.[148]

He has to be describing the clitoris here, surely? Its physical state—the ability to swell and become erect—is explained not in terms of sexual arousal, but in terms of facilitating desire by ventilating the sexually important vagina and cervix.

Some 500 years later, Galen suggested that the clitoral structure provided protection to the entrance of the uterus, in the same way that the uvula protects the throat[149] —again making no connection with sexual pleasure. We've seen how this thinking played out in history.

Evolutionary theory seeks to explain either how a particular trait improves the chances of survival of an organism so that it is in a position to reproduce, or how it aids successful reproduction. Survival and reproductive success mean that advantageous adaptations in an organism are passed to offspring and eventually adapt into the species as a whole. For a comprehensive account of these theories, I relied on Elisabeth Lloyd's book, *The Case of the Female Orgasm: Bias in the Science of Evolution*. It is a rigorous, fascinating, and accessible read. It assesses the past 80 years of surveys about women's orgasms to eliminate any individual survey bias, and gives a comprehensive overview of the theories that have attempted to explain female orgasm in evolutionary terms.

Conception advantages, reproductive benefits

The notion that there is a conception advantage to female orgasm has been popular since man worked out that sperm wasn't carried by the air or gifted by a divine creator. It falls into three categories:

- A woman's orgasm draws sperm into the uterus through the release of oxytocin, which triggers pelvic-muscle contractions that cause the cervix to dip into a pool of semen, and these contractions aid sperm on their journey up the Fallopian tubes. This has been called the upsuck theory and was popularized initially by Fox and Fox in 1970[150] but more recently by Ludwig Wildt and colleagues in 1998.[151]

- Women can control, through orgasm, whether a man's sperm has a better or worse chance of encountering an egg.

- A woman's orgasm stimulates a man's orgasm by the tightening of the pelvic muscles around his engorged penis.

Research into reproductive science has discredited all of these hypotheses, as women's orgasms make not one jot of difference to rates of conception. The first on the list, the upsuck theory, relies partly on missionary-position sex. Today, the missionary position shares equal popularity with two other European and American favorites, cowgirl and doggy-style, which do not create the same seminal pool in the vagina. We don't know the preferred sex position of early man—although it has been suggested that doggy-style would be the most obvious given our evolutionary kinship with apes—and we know from early Christian penitentials that other positions were used despite being frowned upon and worthy of a penance. The conception-advantage theories depend upon coital orgasm being a universal possibility or consequence of intercourse for women, and we know that is not the case. During ejaculation, semen travels at an average speed of 28 mph[152] and is found within the female reproductive tract within seconds, and in the Fallopian tubes within 15 minutes of ejaculation, regardless of sex position or pelvic contractions. The penis is a highly efficient delivery mechanism in its own right.

One study showed that when artificial oxytocin was administered, sperm traveled through the Fallopian tubes faster; but the dose was up to 60 times more than is naturally released during orgasm, so the results don't have any concrete bearing on the hormone's effect under normal orgasmic conditions. It's a bit like saying a turmeric latte helps keep inflammation at bay, when in reality you need to be consuming two-and-a-half teaspoons of turmeric a day, with the right bioavailability agents, for it to be anywhere near effective.[153]

These theories about the role of a woman's orgasm as an aid to conception definitely cast orgasm as necessary in evolutionary terms, and you can see why they are popular. In a world where we hope for female equality, we want female orgasm to be as important as male orgasm. We want to fight the orgasm gap. However, an article in *The Journal of Sexual Medicine* that assessed current research into these ideas deduced that the "bulk of the reported evidence favors the conclusion that the female orgasm, with its concomitant central release of oxytocin, has little or no effective role in the transport of spermatozoa in natural human coitus."[154] This is an unpopular conclusion, particularly with some feminists, because it is seen as diminishing the importance of the clitoris and the female orgasm— perhaps just another conspiracy by scientists to bowdlerize female sexual pleasure. Yes, it does diminish female sexual pleasure when it comes to reproductive sex, but how much of the sex that occurs today is for reproductive purposes? Also, longing for it to have a reproductive purpose does not explain the inefficiency of the clitoris in providing most women with vaginal orgasms. We've seen that anatomical studies have tended to emphasize the vagina over the vulva; for me, this is the feminist issue. Most people don't have sex to practice having babies—they have sex for pleasure (although the issue is more fraught for those with religious beliefs that discourage anything but reproductive sex).

Research has also disproved as bizarre the countertheory, extrapolated from observing female rats and presented by William Bernds and David Barash,[155] that female orgasm has the potential to trigger abortion. The flawed logic by which women were imagined capable of orgasm during unwanted copulation (let's be honest here, rape) to ensure that they do not conceive, defies belief.

Pair-bonding advantages, survival benefits
Historically these have been the most common type of adaptive theories for explaining the evolution of the female orgasm. Elisabeth Lloyd found 19 such theories in this category and cites Desmond Morris's 1967 account as the best known. Broadly speaking, these hypotheses posit that women's orgasms strengthen pair bonds (lasting monogamous relationships

between mating couples serving primarily in the cooperative rearing of young) and that adapting to pair bonds was a significant shift in mankind's evolutionary success. These theories draw on beliefs about why heterosexual pair bonding developed—for example:

- It ensured certainty of paternity.

- It gave all men equal access to women, rather than the dominant men hoarding them. It has been argued that this equality of access promoted the cooperation necessary among early men for settlements to thrive. This concept imagines early men as testosterone-driven, uncooperative brutes appeased by the availability of women, and credits men with driving the development of settlement cultures. Stemming from this is the narrative that women gathered while men hunted—an interpretation challenged by contemporary anthropology.

- It gave women protection while they were pregnant and nursing.

- It provided a secure social environment for raising offspring.

- It makes frequent and regular heterosexual coitus more efficient and therefore more productive. It has also been speculated that this is when face-to-face sex became a thing, as it's a more intimate position. How can we know? Orgasm was one of the factors that "cemented" (Desmond Morris's word) pair bonds, through providing "mutual rewards for the sexual partners."[156]

While pair-bonding theories seek to explain why long-term heterosexual pairing occurred, relying on female orgasm as a binding factor is flawed. It places an enormous emphasis on the social advantage of women's orgasms but does not account for the absence of orgasm for many women in stable heterosexual relationships.

If orgasm gave women a singular reproductive advantage, then surely by now orgasm experienced through coitus would be fully evolved and consistent—or at least more widely distributed among women as an adaptive trait.

Why hasn't coital orgasm evolved for women? An evolutionary case study: What does evolutionary biology tell us about time frames for the assimilation of a trait into the human species? The best example I could find was that given by Elisabeth Lloyd and discussed by evolutionary geneticist Mark Thomas in a segment for NPR on lactose tolerance.[157] The introduction of animal milk into the human diet, roughly 10,000 years ago, gave a natural advantage to those individuals who could best digest the lactose sugar found in milk. Most mammals become intolerant to lactose by adulthood, and this was true for most human populations. However, for some people (including nearly 80 percent of people of European descent), a mutation in the lactase gene which enables tolerance for lactose through adulthood has become common. This is an example of how an adaptive trait has worked its way into the genetic makeup of mankind in what, in evolutionary terms, is relatively recent history. For this reason, the theories about women's orgasms and pair bonding seem wistful rather than rigorous; humans have been reproducing for much longer than they have been consuming animal milk. Possibly those who advance the pair-bonding notions misunderstand the incidence of women's orgasm in coitus and imagine that they are more like men's orgasms—a nearly inevitable outcome of intercourse.

Other theories

For the sake of completeness, there are two more hypotheses about women and orgasm that have circulated in the past. One is that eons ago females (possibly pre-hominid) sought multiple partners until they were sexually satisfied, ensuring that all the males they'd copulated with would protect any ensuing offspring on the grounds that they might be their own. This theory acknowledges that women's orgasms are not quickly or easily achieved with intercourse, although it also assumes they will ultimately be triggered, and it certainly centers women as sexual beings. We have no evidence to prove or disprove it.

Another theory suggests that today's orgasm-in-intercourse discrepancy between men and women (95 percent versus 25 percent reliably orgasm during intercourse) is a recent phenomenon. Ancestral women, it says, were more orgasmic because men had a higher orgasm threshold due to frequency of opportunity. So, in prehistoric times men and women got more sex and more orgasms. Lucky them. One can see that speed of orgasm might be an adaptive trait for men because it would lead to swifter ejaculation, leaving you less vulnerable to attack from those seeking to deter you from reproductive copulation. But this theory doesn't suggest which evolutionary factors drove an adaptation away from orgasm in human females—although having worked on this book it's clear that society has squeezed it out. But did evolutionary biology?

Finally, a hypothesis about the health advantages of the female orgasm. In 1973, Mary Jane Sherfey[158] claimed it had evolved to serve the therapeutic purpose of relieving vaso-congestion in the pelvis. Sherfey suggested that congestion (arousal) ensured that women sought sex until they were orgasmically relieved—but the flaw of course is in the reliability of intercourse to produce female orgasm. Women are not, as Sherfey argues, perfectly adapted for orgasm to occur, because for many the clitoris is not stimulated by thrusting intercourse, so the blood that builds up is not then released by orgasm. You can tell this to the next man that complains about blue, richly oxygenated balls. It's just that clitoral arousal is less visible. What I like about Sherfey's theory is that female orgasm is given a functional benefit advantageous to the woman herself, and it acknowledges female desire and arousal without the pathologizing of hysteria.

Most of these theories treat female sexuality as if it were equivalent to reproductive sexuality, which it's clearly not. Lloyd reminds us of the research that exists on women's orgasms experienced through masturbation,[159] in which 95 percent achieve orgasm in an average of four minutes. Lloyd also points out:

> Gebhard and colleagues note that the "most common masturbation technique is the manual stimulation of the clitoris and the small lips of the vulva," which accounts for 84% of all acts of masturbation among the women the Kinsey team surveyed (Gebhard et al., 1970, p.15). Less than one fifth of women masturbate by inserting an object or fingers into the vagina, and nearly all of those who do accompany the action with clitoral stimulation (Gebhard, 1970, p.16; Kinsey et al., 1953).[160]

Again, so much for the speculum. These findings on the mechanics of how women masturbate have been echoed by subsequent research. What we know supports the view that female sexuality should not be defined solely through the lens of reproductive intercourse, as it has been in the past. This is the lens that has driven evolutionary theory, and perhaps this is why no one has explained the inconsistent nature of orgasm in intercourse in evolutionary terms, or come up with an evolutionary explanation for the existence of orgasm that is not a function of intercourse or produced by mimicking it. This might be a more fertile route for an adaptive theory. Or how about the idea that women's orgasms don't have an evolutionary purpose?

The clitoris is a byproduct.
Lloyd argues that Donald Symons's 1979 evolutionary account[161] of women's orgasms is the most credible. Symons proposed that orgasm in women is not adaptive at all but is a byproduct of embryological development. In the early stages of human embryo development, male and female embryos are only differentiated by the chromosomes they carry, not by any physical differences. At cell level, they are nerve tissues, erectile tissues, and muscle fibers. It is the release of male hormones (triggered by chromosomes carrying this message) at around week eight of gestation that starts differentiation in sexual development. The cells in the female embryo that become a clitoris are the same cells that, in a male, develop into a penis.

Having seen Helen O'Connell's work, which maps the full structure of the clitoris using MRI scans, one can imagine how the same cells would create either a penis or a clitoris, depending on the chromosomal message. They share the same neurological foundation for the reflex that stimulates the muscular contractions of orgasm in each sex. The timing between orgasmic contractions in each sex is 0.8 seconds. (Much popular music taps into this 0.8-second rule. Start with the 1984 Frankie Goes to Hollywood hit "Relax," and go from there.) As a layperson, I find Symons's theory, and Lloyd's advocacy for it, convincing. Fetal cells are multipotential when it comes to sex development, from women at one end of the spectrum to men at the other, with intersex people in between.

This underacknowledged population is as common as those with green eyes or red hair, so human sex is not binary on any level.

Symons postulates that men's ejaculation is tied to orgasm because it is essential to reproductive success, and therefore strong selection on the sexual tissues in men for performance in orgasm, as the mechanism for contractile sperm delivery, is ongoing. Women get the same cells by default, as a happy bonus.

Randy Thornhill and Steven Gangestad point out that Symons's hypothesis only explains the *origin* of the clitoris, and not its persistence as a trait. They also take Lloyd to task for not considering whether female orgasm is *"an evolved adaptation,"* saying that Lloyd "demanded evidence that orgasm is *currently adaptive."* Pursuing this alternative implies that orgasm is newly forming in women, and that maybe there is an adaptive survival benefit rather than a reproductive one. In Thornhill and Gangestad's view, "the case of women's orgasm remains in jury . . . Future research will decide whether women's orgasm . . . indeed does have functional design or, similar to the belly button, lacks it."[162]

Once you understand orgasm in women as a non-adaptive trait, or one that is, as yet, unexplained, it can exist outside of a reproductive function. Dismissing orgasms in women because they have neither an evolutionary nor an (as yet) known purpose is like declining a glass of champagne because the bubbles are pointless. It's like men's nipples—imagine how men would feel if it was declared unseemly or unnecessary ever to stroke or stimulate them. As to women's orgasms, the question becomes, what are you going to do about them? And if you are having heterosexual sex for pleasure rather than as part of the evolutionary thrust of procreation, then orgasm for men is no more necessary than orgasm for women. I'm all for generating a new theory for the evolutionary purposes of the clitoris. How about this: the clitoris evolved to give women their own means of releasing the feel-good hormones dopamine, serotonin, and oxytocin, which also keep the pelvic floor worked out. What's not to like?

Now that we've taken a tour of the history of the clitoris, the next section explores how Western culture has been impacted. Can we link clitoral history to the world that surrounds us? Do past attitudes toward female sexuality and the clitoris explain the patchy knowledge among my daughter's friends and the time it took for me and many of my girlfriends to grow into our sexual selves? Can we see why there is still an orgasm gap in many heterosexual encounters? And what can we do to change this?

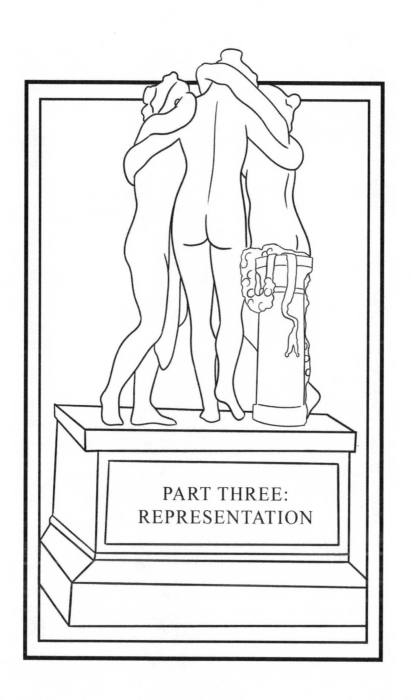

PART THREE:
REPRESENTATION

6
Language

What's in a name?

"I read in a book once that a rose by any other name would smell as sweet, but I've never been able to believe it. I don't believe a rose *would* be as nice if it was called a thistle or a skunk cabbage."

L. M. Montgomery, *Anne of Green Gables*

As we now know, anatomist Realdo Colombo wrote that the part of women's bodies ultimately labeled the *clitoris* should "be called the love or sweetness of Venus,"[163] but sadly his suggestion didn't stick, and I had no such loving term to give my daughter. Instead, I floundered and then used the proper term. What nicknames do we have for the clitoris, and what do they tell us? After all, a character's name is often significant in a story, such as those of Darth Vader, Katniss Everdeen, Jay Gatsby, Voldemort, Mistress Quickly, Scrooge, Roger Chillingworth, and Scout. The list goes on. Nomenclature is important. A name can be more than a name, particularly in a story, and this story is no different. The lack of an easy sobriquet is as telling as the moniker itself.

The origin of *clitoris* has been lost to us. *Penis* has a more literal name derivation, from the Latin for *tail.* This neatly illustrates the difference between the story of the penis and the story of the clitoris. The penis's narrative is less complicated and is easily and widely told. The word most associated with male sexuality is much more frequently used, in all areas of life. When researchers trawled a database referencing over three million academic texts on the behavioral sciences and mental health,[164] dating from the 1800s to the present, they found that the word *penis* was used in 1,482 sources, *vagina* in 409, and the word *clitoris* only came up 83 times. Why is the penis referred to in three times more sources than the vagina and clitoris combined? To what extent are the vagina and the clitoris less relevant to the fields of behavioral science and mental health? The same would be true if we spent hours eavesdropping in public lavatories, libraries, or student common rooms. Both forums, written and spoken, matter—because they indicate to us that the clitoris is still either not considered important, or remains taboo.

Colloquialisms, slang, and pet names

So much for the scientific word. What about the friendly ones? The slang terms that, like nicknames, imply affection? What is the female equivalent of American Dick and English Willy, who might be your homies next door? Anyone with boys in their lives will know how quickly they make friends with their penises. From the moment they learn hand–eye coordination, boys are reaching down for their little acorns every time you take off their diaper. Learning to read, or faced with anything mentally difficult, small (and not-so-small) boys will reach for their penises for comfort. Investigate the pockets of 7-, 8-, or 9-year-old schoolboys and you might find strategically poked holes in the pocket linings. Walk into a den of teenage boys or male students watching a movie, or on a PlayStation, and at least half will be lying on beanbags with a hand down the front of their joggers, protectively curled over their sleeping trouser snakes. Most boys are good friends with their willies; they love them.

List all the slang terms you can think of for a penis. You have three minutes . . .

How many did you get? The National Coalition for Men found 174 synonyms for *penis*.[165] They claim that the penis has more synonyms than any other word in the English language.

Now list all synonyms and slang terms you can think of for a vulva. You have three minutes. Wait a minute you say, this is a book about the clitoris! But for now I'm using the vulva, rather than the clitoris, to highlight the overall paucity of expressions that we have when it comes to a woman's anatomy. If we can't even talk about the vulva, it becomes hard to talk about its specific parts and their functions. If you don't have a map of America, you can't locate or show anyone else where to find the Grand Canyon, Mount Rushmore, or the start or end of Route 66.

List all the slang terms you can think of for a vulva . . .

How many did you get? I struggled to list alternatives, so I canvassed friends in the U.S. and the U.K. to find out what phrases they used with their young children to refer to the vulva. *Vagina* is frequently used, although it is incorrect. Strictly speaking, *vagina* refers to the inner part of a woman's genital tract—not the external part, the vulva. This is where our clitoris's problem lies. If the most-talked-about detail of a woman's body is her vagina, the clitoris remains in the hinterland, an inch or so away from the perceived "important" part, which is the one that enables birth, and is the focus for many men in sex, and the bit that needs dealing with when a girl begins to menstruate. The rest of the vulva becomes irrelevant: unnamed and unused. Indeed, in some circles even the word *vagina* is deemed deeply offensive. In June 2012, Michigan State Representative Lisa Brown was banned from the senate floor for saying it.

Like *vagina,* another misleading phrase is *wee wee,* which is confusing given the proximity of the clitoris to the urethra. It is already a common misconception that they are one and the same. *Front bum* or *bottom* is a commonly used term, but I never liked the idea that one's front bottom is an extension of a back bottom or anus. Just the thought gives me cystitis. *Girl bits* is used in conjunction with *boy bits* to differentiate anatomically. Note that a *bit* is a small piece or part of something that we shouldn't unduly worry ourselves with—but it's a useful blanket term: "Yes, darling, you have girl bits and he has boy bits." Like Lego bricks. We don't name the constituent girl bits, so it's not helpful in our quest for words that allow us to be specific about the anatomy of women; conversely, it's a sure thing that a boy already understands the importance of his penis and the sensitivity of his balls.

The phrase *lady parts* works in a similar way to *girl bits. Lady parts* also implies that they are not something you need to worry about as a child. I believe giving children accurate labels for their anatomy goes a long way to tackling the taboo that surrounds female genitalia and will make adult conversations much easier in the future. An article in *Glamour* (May 15, 2020) reported that a 2019 U.K. YouGov poll found that when asked to point out the clitoris on a diagram of female genitalia, 29 percent of women and 31 percent of men were unable to do so.[166]

Other words that come up are flower terms such as *rosebud* or *lily,* and they explain why losing one's virginity is sometimes called a *deflowering. Foo foo* seems popular. It's the name of poodles in both *The Muppet Show* and the comic book *The Beano,* plus according to the Urban Dictionary it's an adjective used to describe anything that is just a little too fancy. I like this. Fancy is good. Another phrase used is *mini moo,* but that's a bit bovine for me.

Some outdated phrases, such as *tuppence,* were coined by our grandparents. This slang word for the vulva dates way back—my grandmother used it when I was a little girl 50 years ago. Perhaps this is where the old-fashioned British euphemism to *spend a penny* (to urinate) comes from. We could go back to Tudor times, when the whole girl thing was called a *purse.* In the U.K. a purse is a little pouch in which to keep small change. It is definitely feminine in style: think purple velvet or beaded, with a silk lining and a shiny little clasp that shuts with a satisfying snap. Sadly, this euphemism doesn't work where I live now, in America, where a *purse* means a handbag: the term encompasses everything from a small clutch to a bottomless Mary Poppins–style carpetbag in which you can lose your keys, wallet, sunglasses, children's post-school snack, and the dry cleaning. I am sure hospital emergency rooms have stories to tell here—but it's not an image I want associated with my vulva.

Lastly, according to the Online Slang Dictionary, *cookie* is a euphemism for the sexual organs of women. Variations include *cookies* and *cookie jar.* This is an appealing, playful term and the idea that the vulva is a place full of good things is positive imagery at least.

What's striking about this list is the friendliness of the terms—but many of the phrases are babyish. What happens when girls and boys grow up? What words do they use then? When you search *slang for vulva* on the internet, the words that appear are often derogatory and misogynistic. Warning: the analysis you're about to read contains sex and violence. One of my adult sons read this section in an early draft and his response was, "It's a bit offensive, Mum. Don't you think you're going to put people off? Shouldn't you tone it down a bit?" But how many of you have overheard a group of men at a bar or college boys on a night out, talking about sex and the anatomy of women, and haven't been made uncomfortable by the language they used? Some of the words on this list might not be in frequent use—but they are out there. Language communicates and reflects attitudes. Language is not passive; it's about choice, and gives us reference points. These words send us a message about how vulvas are perceived. Let's move from family-friendly nursery terms to the language of more adult slang. Words that girls learning about their sexuality will come up against.

Sorted into categories on the chart below are the words that popped up when I searched *Vulva: what slang words have this meaning?*

Slang Terms for the Vulva

Made for the penis	Unattractive comparison	Unappealing or neutral	More appealing
Bang hole	Hatchet wound	Gash	Candy
Cock pocket	Bearded clam	Snatch	Cookie jar
Prick holder	Beef curtains	Slit	Cherry pie
Fuck hole/slot	Fish taco	Beaver	Cho-cho
Snake den	Ham flap	Minge	Coochie/
Juice wallet	Fur burger	Clunge	Cooter
Sperm bank	Camel's foot	Cunt	Poon/
Sausage mitten	Bat cave	Twat	Poontang/
Sausage holster	Piss flaps	Vadge	Poonani
Hotdog warmer	Soggy box	Growler	Kitty
	Sink box	Bush	Pussy
	Stench trench	Fanny	Smush
	Stinky fahooter	Box	Mitten
	Slobber pocket	Purse	Muff
	Goo cave		Promised
	Prison purse		Land
	Quiet mouth		Vertical
			Smile
			Love box

How do we feel about the vulva now? Many of these words smack of stereotypical and outdated sexist locker-room talk. Only a few of them conjure up attractive images or hint at friendliness or female pleasure. Is it any wonder that this lexicon hasn't been adopted? I found a poignant quote from Maya Angelou about the power of words. Angelou said:

> Words are things. You must be careful, careful about calling people out of their names, using social pejoratives and sexual pejoratives and all that ignorance. Don't do that. Someday we'll be able to measure the power of words. I think they are things. They get on the walls. They get in your wallpaper. They get in your rugs, in your upholstery, and your clothes, and finally in to you.

I worry about the words in the first three columns of my chart and what happens if they get into you and become part of your perception of what a vulva is, or is like. Let's start with *bang hole.* The sex suggested by this term is rough, as when you bang something into place, you whack it in with some force. Also implied in the word *bang* is the idea of a man shooting his load, like bullets from a gun. The vagina, not the vulva, is the focus, and it is presented as a void to be filled by a penis. The vagina is completely defined by the adjective (*bang*) that describes it—and while there is nothing wrong with consensual rough sex, there is everything wrong with a phrase that implies this is the function of the vagina.

The same compound-noun formation is found with *cock pocket,* though at least *pocket* is a more domestic noun: it's a place where you keep things, rather than a void waiting to be filled. Many of the slang terms for vulva imply male ownership; they also reference the vagina, not the vulva, so the clitoris is absent. Thus, our vaginas become handy places to keep penises, and the clitoris is lost—reflecting much of what we found when we explored its anatomical history.

Next up is *hatchet wound,* an extreme version of *gash* or *slash.* This phrase does encompass the whole vulva, but the violence implied by the image makes me deeply uncomfortable on many levels. Why was it caused by a hatchet? Why is my vulva likened to an infliction, or associated with something that has been hurt or damaged? Why would engaging with something like that be anything other than painful or brutal?

Lastly, what do you think about *snatch*? I don't know about yours, but given that my snatch is not for snatching, the word can only add to the confusion about boundaries and what is appropriate sexual behavior. There is a *vast* chasm between the imagery generated by the language we use with young children and the playground or street vernacular that a teenage girl will hear and be expected to use when talking about her vulva.

In the quest to leave the baby speak of *mini moo* and *front bottom* behind, girls and women face an absence of vocabulary when it comes to talking in a more adult way about their sexuality. Men who want to have a conversation with women are likewise at a loss, because so many of the words are misogynistic, violent, or pejorative. There just isn't a lexicon for talking about a woman's vulva, unless one goes for the scientific term.

What's wrong with a scientific term? a) Nothing. b) Everything. Try using the word *penis* around men, and you will quickly find they don't like the clinical sound of it. It's the same with *vulva.* Recently I read an interview[167] about a newly published book on a medieval Welsh poet, a woman who wrote surprisingly bawdy verse about women's bodies and their erotic desire. In *The Works of Gwerful Mechain,* her most notorious poem, "Cywydd y cedor," is translated as "Poem to the Vagina." The translator said *vulva* would have been a more accurate translation than *vagina,* but she rejected it because it had "a slightly clinical ring to it." *Vulva* may be clinical, but choosing the right word makes all the difference to the meaning of a poem where the narrator bemoans men for "ignoring the best bit, silly sod. . . " Raise your hand if you think some men tend to neglect the best bit! Raise your hand if you think some men neglect the vagina! See what I mean about the importance of using genital labels accurately? Otherwise we fall into the trap set by early anatomists of making it all about the vagina. I'm not surprised that Mechain's work was deliberately suppressed by male Welsh scholars in the 19th and early 20th centuries because it was erotic and indecent (caveat: indecent coming from a woman, and perhaps because they felt it emphasized a part of female anatomy that could only bring trouble). Nonetheless I'm sad that an academic translator and editor of erotic poems about the female body, written by a woman, was at a loss for an effective word for *vulva*.

Without a friendly, nonthreatening vocabulary, conversations can't take place. And yes, there are some offensive or at least unattractive phrases for a penis—my son suggested that *cock* was offensive and *hose* made one think of a hosepipe and wasn't very nice either. Men can be protective about what you call their pecker. But the following scenario illustrates just how fraught language around the vulva can be.

The C-word

Imagine you are in a meeting with an antagonistic manager. Which of the following slurs will definitely get you fired rather than a warning?

a) That's a load of bollocks!
b) You're being a prick/dick!
c) You cunt!

I tried hard to think of abusive words or phrases related to men's genital anatomy and couldn't find one that equaled *cunt* in vigor. What does this tell us about the difference in attitudes to male and female sexuality? Why is the oldest slang word of all for a woman's vulva one of the most offensive words you can use in the English language?

There's a cunt story from when I was teaching English. I had a particularly challenging class that included eight or nine disenfranchised 16-year-old boys and a posse of girls. Darren shouted out one day, "Here, Miss, this spelling thing's clever, innit?" Coming from Darren, who was bright and seemed to attend school with the sole objective of disrupting lessons and tripping up teachers, I suspected he was setting a trap. This was the class that proudly told me on my first day with them that their previous English teacher had cried every lesson and they'd never had a teacher last longer than one semester.

"Yes, Darren," I say. "In what way?"

"Take the word *can't*, Miss. Just change one letter and you've got something completely different."

"Yes, Darren," I say, reaching for my marker and writing CAN'T on the board, reading the letters slowly and clearly as I do so. "Which letter were you thinking of changing?" I thought it unlikely he'd say "Swap the *n* for an *r*," and I wasn't wrong.

"The *a*, Miss." Darren is looking smug.

I erase the *a* and the apostrophe, leaving C_NT on the board. I turn to the class.

"What vowel were you thinking of replacing it with?" I ask, pointing to the black space between *c* and *n*. "Say the letters with me: *c* . . . What's that, Darren? What letter did you want to put in here?" I ask again, pointing to the gap.

By now Darren is blushing and refusing to meet my eye, but I don't have to wait long. My reliable wag at the back of the class, Dan, shouts out, "*U*, Miss! He's thinking of the letter *u*."

I write in the *u*. Then I turn back to Darren.

"What does it spell, Darren?" Darren is still blushing and

unusually silent, but George, who can't keep out of anything for long, says, "'Ere, Miss, you can't say that."

"Say what, George?" I ask absentmindedly.

"*Cunt,* Miss, you can't say *cunt.* It's not allowed. I could tell the headmaster on you." I try not to smile. I'm aware there is nothing more alienating than a smug teacher, and I do want to have some success with this class.

"I think you'll find, George, that I didn't say it and you did. But if you'd like to go and talk to the headmaster about this, we could go now, or we could all just get on with the lesson."

The class decided they would prefer to get on with the lesson. It was very productive, and while Darren never let up in his mission to try to humiliate me, there was definitely a shift among some of the other students in my favor. The prospect that I could be so unfazed—so comfortable with the C-word that I didn't blush, skip a beat, or avoid eye contact when talking about it—well and truly did for most of them. Here was a woman who was unlikely to cry in a lesson. I did cry, frequently, but always in the staff room. I also always taught with my classroom door wide open so they could see I was not afraid or ashamed of anything that might happen in our classroom, and I gave senior management a copy of my schedule and invited them to drop in whenever they could as well.

My stance in the face of rude words was made possible by a university friend. Out of his cupid's-bow student lips came vile expletives like dirty kisses. Sitting next to him in psychology and English lectures over a period of three years cured me of any embarrassment surrounding language use, and I can both hear and use obscene language without flinching. This has stood me in very good stead and should be compulsory training for all newly qualified teachers, along with how to use an EpiPen.

The happy button

What about specific terms for our elusive but ever present hippopotamus? A search on *euphemisms for clitoris* comes up with the following short list:

> *Happy button*; *jewel*; *kernel*; *knob*; *little man in the boat* (maybe it's a woman, like God, and Santa Claus? . . . just saying); *love bud* or *button*; *nub*; *panic button*; *pearl*; *skittle*; *sugar plum*; *sweet spot.*[168]

I imagine myself replying to my daughter all those years ago in the bathroom, "Yes, darling, it's called a happy button," or "It's a sugar plum," but I didn't have these words to hand. These slang terms can't yet facilitate a wider discussion about the clitoris and her function because they don't have the circulation or mass presence to make them meaningful.

Purists might argue that men also lack a term for their sweet spot, that *penis* refers to an altogether bigger anatomical structure than the clitoris and is made up of more specific parts responsible for pleasure. However, after some research it appears that no single part of the penis is responsible for sexual pleasure in quite the way of a clitoris.[169] Apparently, pleasurable sensations are experienced all over in varying degrees of intensity (unless you want to talk about the prostate gland, which would be a diversion here). The point is, for women the clitoris is uniquely central to an orgasmic experience. This is why the clitoris matters in the female sexual experience, because for many women no amount of clitoral arousal through the vaginal wall will enable an orgasm, while stimulation of the clitoris's glans is a sure thing.

Call her by her name.

Why haven't widely used and accepted slang words for the clitoris developed in recent history? We have assimilated numerous terms from the world of technology: the smartphone was only invented in 1992, and we're all comfortable with *selfies, emojis,* and *apps.* We might contrast the language we have for the vulva and clitoris with the language that surrounds breasts. They are an altogether different matter. Breasts are universally loved, and the language we use reflects this: *breasts, boobies, balloons; titties, bristols, jugs; fun bags, puppies, melons; baps, racks, bazooms; jams, apples, cupcakes; pillows, sweater kittens, milk wagons; jiggly puffs* and *knockers.* They could be woven into the song "These Are a Few of My Favorite Things." They get lovely euphemisms.

Does the lack of a slang word for the clitoris matter? John Steinbeck's *Of Mice and Men* has been an enduring school study text for years, on both sides of the Atlantic. Ask anyone who has studied it what is significant about Curley's wife and, hopefully, one of the things they'll tell you, apart from the fact she died, is that she doesn't have a name. She is bottom of the hierarchy, low in the pecking order compared with most others on the ranch. And she is Curley's property. Her naming as *Curley's wife* by the ranch hands is driven by the same principles that drive the slang phrases *cock pocket* and *meat wallet* as substitutes for *vulva.* Curley's wife's lack of a name echoes her standing on the ranch: she has none. By this reasoning, the lack of a friendly slang terms does matter, as it's indicative of the level of current discourse that surrounds the clitoris: there isn't much. As with Curley's wife, much language around female sexuality has historically been driven by men. As with Curley's wife, the clitoris doesn't have much status in the world, and hasn't earned a nickname that people feel comfortable with. At least by understanding the history of anatomy and attitudes to female sexuality, we can understand how the vagina has become the locus for discourse about female sexuality—but this needs to change. *Vulva* and *clitoris* should not remain unspoken words.

To use another well-known literary reference, Professor Dumbledore, in *Harry Potter and the Sorcerer's Stone* (or the *Philosopher's Stone* in the U.K.), is quick to reassure Harry that he is right to use Voldemort's name, despite the reluctance of others to do so. Anyone familiar with the series will know that most people avoid using the word *Voldemort* because of their fear of the power of the character.

> "Call him Voldemort, Harry. Always use the proper name for things. Fear of a name increases fear of the thing itself."[170]

Maybe we do need to call a clitoris a clitoris and be done with it. We need to show that we are not frightened of her—there is, after, all nothing to be frightened of. It is the proper name, and eschewing it only furthers the taboo. We know that women who are comfortable using the word *clitoris* are also more likely to climax as part of a sexual encounter. So, whether you use *clitoris, clit, love bud, Sally,* or *hippopotamus,* I urge you to name yours now. And call her by her name, so that we don't gaslight the clitoris but give her a presence through language we're comfortable with. Words can start revolutions.

7
The Written Word

What do books say about the clitoris?

"The more that you read, the more things you will know.
The more that you learn, the more places you'll go."

Dr. Seuss

Understanding the language, or lack of it, when it comes to discussions about the vulva and clitoris provides insight into the anxiety surrounding female sexuality. The language is sparse, and much of it leaves one feeling uncomfortable because it is innately sexist or seems crude. We are stuck in a vicious circle where without the language, conversations are not taking place, and without conversations, the language becomes increasingly remote and irrelevant.

Anne Frank's diary is one of the most famous books in the world. Since it was first published, in June 1947, it has sold 30 million copies and been translated into 70 languages. Who *hasn't* read it? Below is an excerpt from Anne's entry for Friday, March 24, 1944.

Dear Kitty,

   . . . I'd like to ask Peter whether he knows what girls look like down there. I don't think boys are as complicated as girls. You can easily see what boys look like in photographs or pictures of male nudes, but with women it's different. In women the genitals, or whatever they're called, are hidden between their legs. Peter has probably never seen a girl up close. To tell you the truth, neither have I. Boys are a lot easier. How on earth would I go about describing a girl's parts? I can tell from what he said that he doesn't know exactly how it all fits together. He was talking about the "Muttermund" (cervix), but that's on the inside, where you can't see it. Everything's pretty well arranged in us women. Until I was eleven or twelve, I didn't realize there was a second set of labia on the inside, since you couldn't see them. What's even funnier is that I thought urine came out of the clitoris. I asked Mother once what that little bump was, and she said she didn't know. She can really play dumb when she wants to![171]

Don't remember reading this section? Or the next part, where Anne describes in detail "what it all looks like"? You might remember her writing about her period. But this bit? Surely you would remember? Well no, you wouldn't—because it was edited out . . . until the 1995 full Dutch edition and subsequent translations were published. While children and the world have, quite rightly, been encouraged to read the diary, contemplate the terror of being hidden in an attic for more than two years, and confront how such a vibrant, healthy, funny, bright young girl was killed, Anne's unselfconscious, honest appraisal of her body was deemed too much. The world was not ready to read about that.

Shockingly, the world was still not ready to read about it in 1997, when the full English edition was published. The mother of an 11-year-old child in Michigan complained that the unabridged version was pornographic and should not be taught at her daughter's school. But I think Anne's diary would be the perfect way for an 11-year-old girl to start a more mature conversation about her genitalia, and for an 11-year-old boy to learn that there is more to a girl than her vagina. Any sensible teacher would set the entry for Friday, March 24, 1944, as reading homework, pre-empting it with a letter home if their parent body was sensitive, and with a comment to their students that some of the content might be—but shouldn't be—controversial. Any smart school would tie it in with a sex-education lesson. But many copies sold are still abridged versions, and under-resourced school book rooms are full of the earlier editions.

Having learned about the fate of Anne Frank's uncomplicated and candid writing, I was not optimistic, but I decided to trawl moments of orgasm in adult fiction to see what I could find. Then I thought I'd move away from adult references to see if the clitoris came up in teen fiction, and lastly I would review educational texts. I wanted to keep my scope on books that a high-school girl or young woman might come across that would tell her about her clitoris.

So, what might one glean from the classics about sexual pleasure and how to experience it? Gustave Flaubert's 1856 *Madame Bovary* is on many exam-board and curriculum reading lists for English and French. SparkNotes, CliffNotes, and Shmoop cover it, as well as eNotes, so it's a mainstream high-school student read. It's a good book for our purposes, as after its publication Flaubert was put on trial on immorality charges for the "frank and realistic display of the sex, adultery, and other goings-on in bourgeois France."[172] Sounds a promising place to look for the clitoris, doesn't it? I read *Madame Bovary*

expectantly and laughed when I got to the line describing
Emma Bovary's conjugal relations with her husband "at set
times . . . Like a dessert course foreseen in advance, after the
monotony of dinner." Halfway through the novel, Rodolphe
Boulanger de la Huchette is introduced. His surname translates
as "baker of Huchette," meaning someone who makes and sells
cakes, pastries, and tarts—in other words, desserts. Given the
earlier simile linking dessert to coitus, his is a prophetic name
if ever there was one. The handsome visitor suggests to Emma
Bovary's husband that "riding" would be good for Emma's
health and soon thereafter Rodolphe turns up uninvited at the
Bovarys' front door, with two horses.

Next thing we know, Rodolphe and Emma are galloping
through the countryside side by side with their knees jostling,
horses panting, and the leather saddlery creaking with the exer-
tion. Dismounting, Emma and Rodolphe tie up their horses for
a rest and together they walk toward the woods. Emma lifts her
riding habit up to expose her fine white stockings, and her veil
slips as she clambers over fallen branches and tussocks. (You
tease, Emma.) Rodolphe puts his arm around Emma's waist and
draws her toward a marshy pool. Soon, Emma is leaning in to
Rodolphe, throwing back her neck and, we are told, "swelling
with a sigh, and . . . with a long shudder and hiding her face,
she gave herself up to him."[173]

Any reader of romance fiction knows that "giving oneself
up" is a euphemism for having sex. Emma "felt her heart, whose
beating had begun again, and the blood coursing through her
flesh like a stream of milk. Then far away, beyond the wood,
on the other hills, she heard a vague prolonged cry . . . mingling
like music with the last pulsations of her throbbing nerves." The
description is over in a short paragraph but, by the sound of it,
it was a successful quickie. Surely the "pulsations of her throbbing nerves" and the disembodied cry were those of Emma's

orgasm. Surely. Or was it just a particularly good pastry? And why does "marshy pool" make me think WAP?

Madame Bovary is clearly a book with some great sex. When Emma gets to her second lover, Léon, in a bateau bed surrounded by red velvet curtains, "the great knobs on the andirons would gleam suddenly," and on the mantelpiece next to the two candlesticks there are two "large pink shells." The sex is off the scale but, to the sexually inexperienced or a reader not primed to be looking for double meanings in the description of the fireplace and surroundings, it's gobbledygook. You could read it to an innocent young teen and she would be none the wiser. And that's my point. Where's a girl to find out the actual mechanics of sex and how you arrive at pulsations and disembodied cries, with descriptions of the metal frame within a fireplace, candlesticks, and shells to go on?

I imagine that Emma Bovary time travels and I am fortunate to get a candid interview about her sexual relations.

Me: So, Emma, Rodolphe was good then?

Emma: Yeah. He knew what he was doing—he's been around after all—but I gave myself a bit of help.

Me: How did this go down?

Emma: Oh, he really liked it. It seemed to excite him. A bit too much, actually; he was over sooner than I would have liked, but I was able to make the best of it. Mind you, he'd lit a cigar by the time I was the other side of it, congratulating himself on how well it had all gone.

Me: (*Laughing.*) Nothing changes, then. The postcoital cigarette was literary and cinematic code for "It's over and I was good" right up until smoking got banned on public transport. Tell me a bit about your relationship with your husband, Emma.

Emma: Victorian. You know ... paranoid about women being too sexual. He read some book by Rousseau[174] and that blessed Sophie girl who's *so* perfect and passive and all, into sewing and housework and not at all interested in sex, and my sexual fulfilment was off his radar, like it is for every other decent woman in France. That, combined with his mother and her views on women ... the cow. The word *bovine*'s too good for her. She's aptly named, don't you think? Bovary, Bovine, they even sound the same. Charles, he likes me coy, so I play along. He makes out he climbs on top of me for missionary-position sex out of duty, but I know he likes it from the way his fat fingers fumble with his trouser buttons when he's getting his todger out.

Me: I don't want to be rude or anything, but your adultery does feed into Rousseau's commentary on the morally fragile nature of women and the male paranoia about women's sexuality, doesn't it?

Emma: (*Laughing.*) You seem to have forgotten that I am entirely a figment of a male imagination! I was written by a man. Haven't you noticed that so few novels of this era feature adulterous men? Do you really think this was because there weren't any? It's laughable, isn't it? It takes two to tango. I mean, adulterous women are probably doing it with adulterous men, but nobody writes about them.

Me: You seemed pretty cut up when Rodolphe, your first lover, didn't follow through with your joint plans to run away together.

Emma: I was! I mean, no girl likes to be jilted, but also I thought I was going to get away from all that open countryside and I was *so* excited about going abroad to Genoa ... That was the real disappointment. I'd gone to all the trouble and expense with the luggage and packing. (*Pause.*) I'd have got bored of him eventually, though; you know me. And if he hadn't left, I wouldn't have got together with Léon.

Me: Yes, tell me about Léon.

Emma: Now he *was* good! (*Emma flushes and pauses, remembering.*) Twice in a row, straight up like the candlesticks on the mantlepiece. But all that nonsense with the andirons and the shells—why on earth couldn't old Gustave have just said it outright, throbbing you-know-what and vulva? Léon, he had a beautiful candlestick, and who'd have thought it but he knew what to do with a shell—I felt the waves rushing over me.

Me: Thank you, Emma.

The issue, apart from the fact that the sex is written about in code, is that a lot of 18th- and 19th-century literary heroines who have great sex are creations of 18th- or 19th-century male imaginations, and it doesn't end well for the fictional sexualized women. Take Anna Karenina, for example. As if those symbolic thundering trains thrusting through the novel were not enough, she has to die underneath the wheels of one, too. Now I know about Victorians and trains, I get that it's a metaphor; she gave into sexual passion and it killed her.

Most male writers of the era gave their sexualized women endings that were endorsed by the establishment, as defined by society and the Church at the time: sex was unbecoming for a woman, should be reserved for the procreation of children within marriage, and any sex outside of that resulted in punishment, be it social exclusion and family shame, or death. The death could either be real or a metaphorical social death, so they died either way. R.I.P., Tess of the d'Urbervilles, Anna Karenina, Becky Sharpe, and Hester Prynne. R.I.P., you flushed heroines of gorgeous 19th-century operas. R.I.P., you maids of ballads. Great sex in an Enlightenment, Romantic, or Victorian novel is unenlightening because the clitoris and orgasm are referred to only metaphorically, and the sexist sexual mores of the time are reinforced. The doomed heroines' lovers go on to other mistresses and more sex, with marriages to suitable women and successful lives. The women, on the other hand, die, become prostitutes, or languish in poverty and isolation. That is, apart from Fanny Hill, although her story is problematic in other ways.

Fanny Hill is the eponymous star of the fictional, much-banned *Memoirs of a Woman of Pleasure* (1748).[175] She is definitely not *literary* fiction; she is gold-foil print all over the cover. Fanny was freed from the moral guidance of parents, like many heroes and heroines in books, by becoming a young orphan. She quickly discovers her love of sex in a brothel and embarks on a five-year sex fest. From the safety of her marriage, significantly to the man who first deflowered her in the brothel, Fanny recalls her escapades in two intimate letters. The young Fanny witnesses and experiences *a lot* of orgasmic sex, all with great glee. Her first orgasm comes though masturbation, while witnessing two other couples in flagrante from a closet—although she had come close to orgasm at the hands of Phoebe in a lesbian encounter prior to this. Later in the novel, a benefactor proves to be a very satisfactory lover, and while much of the sex exalts large penises and vaginal orgasms, Mr. Norbert also lavishes attention on Fanny's clitoris (although the clitoris itself isn't named, the writer preferring to refer to it as a "secret and critical" part):

> Kissing me in every part, the most secret and critical one so far from excepted that it received most of that branch of homage. Then his touches were so exquisitely wanton, so luxuriously diffused and penetrative at times, that he made me perfectly rage with titillating fires.[176]

Fanny was the dreamchild of John Cleland, his get-out-of-debtor's-prison ticket. Cleland enjoyed a year of publishing success before he was forced to denounce his book, and Fanny, on the grounds that it was "corrupting the King's subjects." However, the book was pirated, reprinted, banned, reprinted, smuggled, reprinted, illustrated, illustrated again, and again, and circulated for centuries on the black market in both the U.K. and the U.S.A. It finally achieved the status of legitimate print runs more than 200 years later, in the 1970s, and was still a crowd-pleaser. You might not have heard of Fanny, but your mother or grandmother definitely will have.

Memoirs of a Woman of Pleasure is erotic, but though it's historically interesting for the sheer fact that it details female sexual pleasure with such openness and glee, it is also problematic for a contemporary audience. Fanny turns 15 as she starts her five-year sexual foray—definitely underage. The text romanticizes prostitution and endorses scenes that are arguably rape. It is contradictory in that it has no problem with girl-on-girl scenes, but is damningly judgmental about a gay male encounter. And for the purposes of this book, where I want to look at texts a teenager might encounter, and what they might learn from them about female sexuality and the clitoris, *Fanny Hill* is an unlikely contender.

After successfully interviewing Madame Bovary, I fantasize about inviting to tea a group of 18th- and 19th-century female authors with mainstream acceptance, to ask them about their craft and how they approached writing about sex. In my head, Aphra Behn and Mary Wollstonecraft are the first to arrive. Both women were vilified and discredited for their unconventional private lives. (That's code for "They had affairs," which was, as we know, O.K. for men but not for women.) They toss their hats onto the sofa and their talk is animated as they compare their experiences of being written off as morally depraved.

"Really, libertine!" says Behn. "I was, it's true. But then so was every other Tom, Dick, and Charles of the time. Right royally so—look at Charles II and the way he flaunted his mistress Nell Gwyn and their child. But I was a woman, so the world disapproved. Although every heterosexual male libertine needs a female one to consort with—idiots."

Mary nods sympathetically.

"I know. The hypocrisy," she says.

At this moment, a group of men arrive, and I open the door, chain on, to tell them to go away.

"Let us in, quick!" one of them says, before I can shut the door. "We had to come in disguise."

George Sand, George Eliot, and Ellis Bell step into my hallway, looking for the table to drop their calling cards on. All three found it advantageous to publish using male pseudonyms. George Eliot and Ellis Bell discard their hats and cloaks, pull out their dress skirts from the tops of their pants, and kick off heavy boots to reveal Mary Anne Evans (whose books, most famously *Middlemarch* and *The Mill on the Floss*, still bear the pseudonym George Eliot) and Emily Brontë (a.k.a. Ellis Bell). But George Sand, author of more than 70 popular French novels, who frequently dressed as a man as well as using a male name, keeps her male attire on.

"The train from Nuneaton was delayed, I had hoped to be early," apologizes Evans. "And please call me George. Everyone else does."

"Yes, the Leeds train was late as well. We met at the station; wasn't that fortunate?" explains Brontë. "This industrialization isn't all it's cracked up to be."

There is one more knock at the door.

"It's me, 'A Lady,' a.k.a. Jane Austen!"

I invite my guests[177] to sit down at the table and I pass around the cucumber sandwiches as I explain my mission: to unravel attitudes to the clitoris in current society by exploring the past. Silence falls.

"That's your thing, Aphra," says Wollstonecraft. "By my day every woman was labeled as weak willed, irrational, and sensual. Bah! What do you expect if you don't educate women? There was so much sexual inequality! I tried to vindicate women, but my romantic affairs didn't help . . . I was, as you would say, slut shamed. It is such a joy to me that now I'm being read for my ideas, rather than dismissed for my irregular relationships."

Behn, who, like Wollstonecraft, struggled with poverty at times during her career as a result of her sexual choices, has finished her sandwiches and is ready to speak.

"Yes, I was a proponent of sexual freedom . . . but not a successful one. Good luck with that. Mind if I have a cupcake? Hunger is a terrible thing for creativity."

I ask the group about how they framed sex, particularly from a woman's perspective, in their novels.

"The world wanted me, as a spinster, to blush at the thought," says Austen. "I wasn't supposed to know much about it, and really, everyone was so obsessed with marriage, that's all they wanted to read about—daughters finding matches. People weren't interested in what happened next. *Of course* it was going on. There were Willoughbys and Wickhams aplenty, your modern-day players. In my time, they tended to be portrayed as soldiers with dashing red stripes down their uniform trousers, because aside from uniform being a turn-on, red signaled danger. Soldiers were also associated with risk and fast living because of the nature of their lives—everything about them signaled trouble and excitement. It didn't do for a girl to get too involved. There wasn't a future in it unless you were already married, and certainly it wouldn't have been seemly for me to write about that. It was more than my tedious family would have been able to bear, given my spinster status.

"You could try Gothic fiction, that was awfully popular in my day. Awful being the operative word. It was all cloak, dagger, and guttering-candle stuff. You know, a pale maiden finds herself inured in a gloomy castle on some Eastern European mountain by an exotic man. But readers still had to fill in the gaps for themselves. Young girls imagined being kissed by red-lipped counts and couldn't explain why it made them feel so tingly and fidgety, and married women took it as far as they wanted, in the safe realms of their heads. It was a mix of horror, terror, taboo . . . and subliminal sex."

"Besides," says Brontë, "everyone was *so* judgy and edgy about sex, and women and sex. It was hard enough getting published, even under a male pseudonym. Although really that's what it was about with Catherine, Isabella, and Heathcliff in *Wuthering Heights*," she continues. "The sex. Kinky, Heathcliff kind of sex. He met his match with Catherine, but Isabella was a good submissive. Do you have a copy of that book *Fifty Shades of Grey?* I'd love to read it. I gather it's been quite popular."

"I have, actually," I reply. "All three. You're very welcome to them."

The *Fifty Shades* series has been hugely popular, selling more than 125 million copies in 52 languages. My daughter (then age 18) offered to read them for me as part of my research, promising to mark the sex scenes and references to the clitoris. After a couple of days, she brought them to me, the second book unfinished.

"Didn't you like them?" I asked.

"No. They made me angry. The main character, he just owns her sexuality. He's so controlling, and the sex is just a symptom of that. The clitoris does get mentioned, mostly licked, and it's all about what he does to it, which I guess would be good if you liked the guy. I'm not going to finish it. I'm fed up with this older man–younger girl thing. It's creepy."

That is why the first book didn't appeal to me when it was published. I love that E. L. James (another writer who found it expedient to have a gender-vague pen name) made a hit out of writing about sex, I just wished she'd been more feminist about it, maybe cast an older woman and a younger man, or challenged the "broken man gets fixed by a good woman" narrative. It's a plot trope that doesn't lead to healthy relationships for women.

In a TED Talk about unwanted arousal, Emily Nagoski (writer, educator, researcher, sex therapist, activist) points out that the female protagonist of the *Fifty Shades* series, Anastasia, clearly says she doesn't like being tied up and left hanging, yet the "hero" contradicts her, ignoring her words and writing his own narrative about what she wants. Like Nagoski, I don't care if you go in for S&M, but I do care that it brings you pleasure. And would I be old-fashioned in advocating that young women practice vanilla sex first, until they get familiar with it?

It is unlikely that a teenage girl is going to be given a *Fifty Shades* title for her birthday or come across one on her school syllabus, although the books may be passed from one student to another surreptitiously. If you discover this is the case, please have a conversation about consent, about not doing things just to please a partner, and about her right to pleasure.

The truth is, liberal as I am, I don't enjoy reading explicit sex scenes. They mostly make me feel uncomfortable. If I want something steamy, I'll watch an episode of *Outlander*. The point is not that fiction should contain more sex; but if as a girl or young woman you're not finding out about it through your reading, then where are you going to get your information?

Teens and pornography

The trouble with teen fiction is that it's often about first-time experiences and, in reality, how often are they orgasmic? It also tends to be written in such a way that it doesn't tell you more than you already know. It's vague, and a reader is left filling in the gaps, which is why we need to be giving teens accurate factual information. The place, surely, for good information about your clitoris, if you don't get it from your mother, aunt, or grandmother, from classic literature or your science book, is in a book specially designed for you, about you, given to you to read at the right time, somewhere between ages 10 and 15. In a perfect world, boys would also have access to this information. Men on the whole want women to enjoy sex as much as they do, and most teenage boys are definitely hungry for any information they can get about girls and sex.

What about pornography? Some men fondly remember their first porn-mag heroine. Ask them. Many boys get a lot of information about sex from porn. But girls? Often girls feel anxious about porn, because they know it can be exploitative, and in the back of their heads there is a voice telling them it's shameful. A foray into porn quickly turns up a host of images and movies designed for the male imagination. We've had that before, haven't we, the male imagination? It is a recurring theme.

Here are some stats on teen usage of porn. I know that I digress here and this is moving out of the realm of print to include film, but it is relevant to our conversation because it shows the discrepancy between boys and girls, and the role porn plays in young people's sex lives.

- 93% of boys and 62% of girls are exposed to internet porn before the age of 18.

- Only 3% of boys and 17% of girls have never seen internet pornography.[178]

- In a study of 304 random porn scenes, nearly 90% contained physical aggression toward women. Close to half contained verbal humiliation. The women nearly always responded neutrally or with pleasure.[179]

- 87% of college-age men had viewed porn in the previous year, compared to 31% of college-age women, with slightly less than half of male college students using it weekly compared to 3% of female students.[180]

- MindGeek (which owns PornHub, Brazzers, YouPorn, and Reality Kings) is one of the top three bandwidth-consuming companies in the world, alongside Google and Netflix.

- Most porn is consumed on smartphones.

- Teens say they use porn to learn about sex.

- Most parents believe their pre-teens and young teens are not watching porn yet and say they have discussed porn with their kids.

- Most teens have watched porn and say there hasn't been a significant conversation about it in their lives.[181]

Boys do consume more porn than girls. Is this because boys like porn more? Or do they feel pressure to watch it? Or less stigma about watching it? Or when they watch it, do they find that it speaks to their needs, while girls do not find it as appealing? Or does the exploitation of women in the sex trade deter girls from endorsing it by watching it? I don't know the answers, but I feel strongly that porn is not the place we should be sending our children for information about sex and their sexuality. Just as we wouldn't send them into a sex shop. Porn is not going to go away, so we need to counterbalance it.

Sex-education books

My daughter and I embark on a trip to our closest Barnes & Noble, a big store with a café and friendly staff, to search the children's and young adults' sections. We are looking for the word *clitoris,* and want to learn how female sexuality is dealt with in current sex-education books. In the children's reference section is a series of color-coded, age-appropriate hardcover American Heritage dictionaries; the vulva and clitoris are added to the *Student Dictionary.* A student looking up *clitoris* will find this definition:

> **Clitoris** *n.* A sex organ that is composed of erectile tissue and forms part of the external reproductive system in female mammals and some other animals.

Although scientifically accurate, this is singularly unhelpful in terms of what a clitoris might do for a human vulva owner—as opposed to a female mammal—but I can see that by keeping the focus on mammals in general, the topic of human sexuality is avoided. The syntax in this definition is interesting because the femaleness is held back. Why doesn't it say, "*The female* sex organ," which would acknowledge women as sexual beings? The penis definition, after all, says:

Penis *n.* 1. The sex organ of the males of mammals, reptiles and certain birds. Most male mammals also use the penis for urination.

The word choices in these definitions reveal that male sexuality is prioritized and endorsed and female sexuality is obfuscated.

Right away, the penis is branded as male: The definite article ("the") is used to precede a specific noun, whereas the clitoris description gets "a," an indefinite article, used to refer to things in a less specific manner. It implies one of a number of important parts. Which would you rather be, *the* girlfriend, or *a* girlfriend?

The penis is, actually, one of 14 sex organs: For example, men also have a scrotum with testes, and an essential sperm highway in the vas deferens is kept flowing by the seminal vesicle.

A woman's reproductive system is divided: It has external and, by implication, internal parts, as if the "external" bits aren't crucial for women. The penis is never labeled as "external," but it is, isn't it? And it is entirely inaccurate to say that the clitoris is external, as the external part is only the tip of the lovely erectile whole. Books published in the last century might be excused this misconception, but anything published in the past 15 years should not be.

Having taken issue with the word choice, I also have a problem with the sentence structure in both definitions because it "bigs up" the penis and diminishes the clitoris. (All very Freudian.) Count the words between "The sex organ" and "males" in the penis definition. Two. It is unambiguous. Don't we know it. Now count the words between "sex organ" and "female mammals" in the definition of the clitoris. Fifteen, in a compound sentence that banishes woman's ownership of a sex organ to the end of an altogether more complicated sentence, like this fact is being hidden. Why can't the clitoris have a simple sentence too? I suggest instead:

Clitoris *n.* The sex organ of the females of mammals and some other animals.

Does syntax matter? Am I being picky or does the andro-centric model that drove science for so many years continue to permeate these definitions? Ask any high-school junior if syntax matters, and they will tell you it matters if you want to do well in the language section of the SATs or ACTs. Syntax is essential to meaning. It also creates nuance and emphasis. Syntax is the difference between *extra marital* sex (meaning more, like extra chocolate sauce on your home-baked cheesecake) and *extramarital* sex (meaning occurring outside of marriage, with another person, like going out to get another portion of cheesecake in secret; it's a whole different deal).

The language used in the American Heritage dictionary series is subtly sexist, and I'm sure this is unintentional. Here lies the problem: the inherent sexism wasn't intended, and no one challenged it, because the difference in attitudes to male and female sexuality have become so embedded in American and European society that this quietly pervasive prejudice continues to creep through our lives. It happens at such a subconscious level that anyone who fusses about it lays themselves open to accusations of being extreme or oversensitive. Really? Quibbling over whether it's a definite or an indefinite article? Who cares?

We should care about language. Quibble more. It's our language that defines us. The clitoris is *the* organ of sexual pleasure for women.

Back to our Barnes & Noble visit. A sales assistant pulls out some books for us: a purple hardback, designed for eight- to ten-year-olds and their parents and caregivers, called *Sex Is a Funny Word*, and a larger paperback book called *It's So Amazing!*

"And these are really popular for young teens," she says, handing us a pair of slim books: a blue and green one—*What's Going On Down There? A Boy's Guide to Growing Up*—and a paler book with pink design details, *The Period Book: A Girl's Guide to Growing Up*. We take our pile of books to the café so that we can see what we're getting before we purchase.

The *Sex Is a Funny Word* hardback makes us smile. The drawings are cheerful and free from gender stereotyping. The information is unfussy and clear. We like the way there are pages emphasizing respect, trust, and justice. The balance of information seems scrupulously fair and it is refreshing not to have the section about a girl's genitals in baby pink. The confusion between the vulva and vagina is acknowledged, and the clitoris gets a proper write-up: "The clitoris can be very sensitive, and touching it can feel warm and tingly." Right away the clitoris is acknowledged as a source of pleasurable sensations, and the anatomical description is detailed and accurate, telling me more about the clitoris than I knew until I started researching this book. Hooray! Parity. This is great news for female sexuality; a generation of educated girls, boys, and non-binary children who know that a clitoris exists. Under the heading "Erections," we're even told that the clitoris can get erect.[182] The equality is fabulous, and the simple factual statements about pleasure refreshing. Just when we think this book can't get any better, we turn the page and there's a section that introduces the word *masturbation.*

> Touching yourself is one way to learn about yourself and your body, and your feelings. You may have discovered that touching some parts of your body, especially the middle parts, can make you feel

warm and tingly. Grown-ups call this kind of touch masturbation.

There is a complete lack of judgment or stigma. The learned men of science of the 19th century will be turning in their graves. This book is so accessible and cheerful, educating children about themselves and their gender differences and sexuality in an uncomplicated, equitable way. It's the book I needed all those years ago to read with my children after the bath.

Imagining the learning journey that today's well-informed, unembarrassed, and curious young readers will take, we turn to *It's So Amazing!* This heteronormative book focuses on reproduction, and it's good for understanding where babies come from. The *amazing* bit in the title refers to pregnancy, not sex; there is very little about actual sex. The clitoris is shown as a dot on a diagram and gets the briefest mention as "a small bump of skin about the size of a pea." With information like this, no wonder it can take years to discover one's clitoris.

Toward the end of the book, masturbation is covered, with an acknowledgment that different families and religions have their own "thoughts and feelings about masturbation," but it concludes with the guidance, "Most doctors agree that masturbation is perfectly healthy and perfectly normal—and cannot hurt you or your body." The stigma often associated with masturbation is neutralized and presented as a gender-neutral activity, which is good, otherwise some girls grow up thinking masturbation is, like farting, exclusively male. But how's a girl to know how to do it? She could easily assume it's all about inserting something into her vagina and, after finding that unsatisfactory, give up, hoping that the magic of a penis will provide the orgasmic experience she's witnessed happening on TV.

M y daughter and I purchase our books and return home to peruse the two early-teen guides to growing up. We wonder why it's necessary to have separate books. Gender separation in sex-education classes in schools at this age may make sense, because it minimizes embarrassment and stops the sessions from being hijacked by single-interest groups (although I think it marginalizes intersex and trans teens and limits knowledge), but in a book designed to be read alone and then possibly followed up with a discussion at an appropriate time with a parent, where's the need for separate information? Surely it's helpful for everyone to know what's going on with each other's bodies? Most people will have close friends and siblings who are not of their biological sex, and will likely, one day, be parents to children of other genders.

We're curious to see what aspects of female sexuality are covered in *The Period Book: A Girl's Guide to Growing Up*. Skimming through the index it is noticeable that sex, orgasm, and masturbation are absent. The clitoris is referenced twice— once as a dot on a diagram of a vulva (although this word isn't used anywhere in the book), and again in a three-sentence description that tells us the clitoris is:

a) Not an opening
b) A little button-like bulge
c) Sensitive and results in pleasurable, sexy feelings when touched.[183]

That's it. The closest thing to any sex talk is a chapter on "Romantic Feelings," which is defined as *having a crush* and comes near the end of the book, before the chapter on "Sexual Harassment." It strikes me we haven't moved on from the tingly feelings of the book designed for eight-year-olds.

"I'm disappointed," my daughter says. "And really fed up with 'romantic feelings.' It's exhausting! When do I get sex or orgasm? And where's my clitoris? This book would be good if you were eight and wanted to know about getting your period. It tells you a lot about that. But it doesn't tell you *anything* about sex."

This is not a sex-education book. It's a comprehensive puberty and period book, with very little content about what becoming a physically mature female means sexually. As *A Girl's Guide to Growing Up* it covers body changes with puberty, bra styles, your period, and period anxieties, whether you should shave your legs or not, how to stay fresh, puberty "bummers" like spots, tender breasts, cramps and mood swings, body image, changing friendships, and sexual harassment. A girl is reassured that it is O.K. to think about a boy (or boys) she really likes, as long as she doesn't let it take over her whole life. It seems she'll be so anxious about leaking, discharge, staying fresh, and being at her best weight that she won't have much time for thinking about boys anyway. Her clitoris has been reduced to a pea, and she is awash with "sexy" and "romantic" feelings that are so vague she will have no idea what to do with them. Any parent who buys this book to help their daughter learn about being a sexually mature female is misguided. Woefully so.

It's alarming, given that only 43 percent of parents say they feel very comfortable talking with their children about sex and sexual health. This means 57 percent only feel somewhat comfortable, or downright uncomfortable.[184] Discussions should be supported with good books. Pregnancy, STDs, contraception, and making the decision to have sex are the main focus of sex-education lessons in school and tend to be the topics most discussed at home. In a U.S. survey of 15- to 17-year-olds about why they may not be able to talk to their parents about sexual health issues, 83 percent of girls said they didn't know how to bring the subject up. So if you as parents don't tell your daughters about the clitoris and sexual pleasure, it seems probable that most young women today won't ask *you* about it.[185] They'll probably ask Google instead.

Some people think this information can wait. However, according to 2015 research, 41 percent of American high-school students reported having had sexual intercourse and 30 percent reported being sexually active, meaning they'd had sex within the past three months.[186] By 19, three-quarters of American teens have had sex. The statistics are similar for the U.K. By high school, girls need to know about sex. Given that nearly half of them are having it, wouldn't it be better if they also knew how to have sex that was pleasurable? Orgasmic, even?

If you really want to understand what the sexual landscape looks like for today's teens, read Peggy Orenstein's 2016 *Girls and Sex*.[187] You may not like what you learn, but forewarned is forearmed. One argument against comprehensive sex education is that it endorses and encourages underage sex. This is not true. Research firmly shows that education delays or reduces sexual activity, and certainly increases the use of condoms and other contraceptives.[188]

We need to be talking to girls about all aspects of sex and their sexuality, not just about reproduction, contraception, STDs, and virginity. The evidence tells us that the more information a girl has about sex, the longer she waits to have it. Sex becomes less of a mystery, less of a biological function she wants to experience or to provide for boys; it becomes something that's better done with a partner she knows and trusts. For some girls, their virginity hangs less heavy on them. The definition of *healthy sex* needs to be expanded so that it doesn't just mean sex free of pregnancy, chlamydia, or herpes; it should encompass healthy, satisfying sex, both emotionally and physically.

For most men, the end result of sex, be it sex play or intercourse, is orgasm. Let's educate girls so that their chance of orgasmic sex play is similar to that of boys, and then they might feel more empowered to wait for intercourse, if that's what they want. They might also have better sexual experiences overall. Most girls are so unaware of their own sexuality that they believe intercourse *is* the route to an orgasm. If we can't talk about the clitoris, masturbation, and orgasm with the girls in our care, we should ensure that they have access to good information from other sources, such as books. Why would we wait for them to find it out for themselves, if they are lucky, from a college roommate, a more experienced lover, or *Cosmopolitan* magazine?

We wonder what the companion book to *The Period Book* covers, and my daughter picks up *What's Going On Down There? A Boy's Guide to Growing Up.*[189] She flips to the index.

"Boys get orgasm *and* masturbation!" she says. "They get orgasm three times. And masturbation three times! *And* sex. Lots of sex: sex play, sexual intercourse, sexual performance, sexual stimulation!" She is outraged. "They get a whole fucking chapter on having sex!" Turning to it, she reads the start of the chapter aloud: "Since having sex is a topic that many boys are *very* interested in, some of you may have turned to this chapter right off the bat." It proceeds to tells boys how to have sex, how the vagina accommodates the penis, how it produces fluid to make it slippery, and how it's elastic so the penis will fit in. It doesn't tell you much about the clitoris, although it does say it's responsible for many of the pleasurable feelings women experience when they have sex—but, as my daughter points out, "a pleasurable feeling's not an orgasm though, is it!" This is no better than Krafft-Ebing in 1886.

"Do the boys get romantic feelings?" I ask. She turns to the contents page.

"They get 'Stress, Pressures, and Weird Emotions.' Does that count? Hey, listen to this. It says, 'No question about it—getting older has some definite perks!' I can't find any romantic feelings," she says, "although I do see 'being sexual.'"

I'm reminded of the gap in expectations that I've seen played out in teenage relationships on school campuses, on both sides of the Atlantic, as well as in high-school movies, where the girl is focusing on the fuzzy romantic aspect of having a boyfriend while the boy is focusing on getting laid. To what extent is this driven by evolutionary, biological needs, rather than social expectations and attitudes? If girls are handed romantic feelings as the definition for what they will feel about boys, and boys are given excitement about sex, no wonder high-school and college encounters frequently fall short, with both sides feeling let down and confused.

Yes, girls and ladies, you're paid less and—now you know—you're worth less sexually. Your orgasms don't count, yet boys have theirs sanctioned with chapters on wet dreams, masturbation, and sex, all the while being reassured that these experiences are absolutely "normal and natural." You're much more likely to have an erection, boys are told, when "stroking or rubbing your penis" or "looking at a girl's breasts." How come girls aren't given advice on how to become aroused? What happened between *Sex Is a Funny Word* and these books? Boys are told that when men are asked to describe how an orgasm feels they say things like, "fabulous, exciting, wonderful, or even the best thing ever." Yes, please, I'll have some of that! But oh, I'm a girl. It's not on offer, I forgot. The boys' book even advises:

In fact, masturbation probably helps prepare you for sex with another person. By exploring your own body, you have a chance to learn what feels best to you. Later you'll be in a better position to let your sexual partner know the things that you like.

So not only do boys get excitement, information, *and* orgasms, they are also encouraged to find out how they like it and ask for it that way. Compare this to the content in the girls' companion book. Both books are "bestselling," and "updated"—in 2017. Did we really feel even in 2017 that boys and girls should be treated differently when it comes to sexual knowledge and expectations? Shouldn't we be striving for sex equality, as well as gender equality? This raises an uncomfortable question for many whose gut preference is for girls to be naive and chaste when it comes to their sexuality, yet who acknowledge how inherently unfair it is. This gut response needs to be challenged. It is a vestige of the flawed thinking that shaped much early science and psychological theorizing.

I'm so confused by the disparity between the two teen books that I examine them more closely. For the girls' book, the author and illustrator asked girls aged 8 to 12 what questions and concerns they had about puberty. For the boys' book, the author thanks "men and boys who shared their thoughts and experiences about puberty." The age differences might reflect the fact that girls tend to reach puberty at a slightly younger age. However, the books are definitely designed as a pair: they share the same subtitle, typography, color palette, author, and publisher. I am struck by the way boys and men are clearly not afraid to ask for the nitty-gritty about sex, such as how an orgasm feels and how you might get to have one. In contrast, the girls' questions, comments, and observations didn't yield the same content. I am reminded of a lecture I attended on market research. The speaker's opening comment was:

> The first rule of market research is, don't ask people what they want. If you ask people what they want, they will only be able to play back to you what they already know. The real value of market research is teasing out what your target market don't know

they need yet, and then discerning how to give your concept traction, how to make it fill that untapped need as effectively as possible.

Girls and young women want information about sex and their sexuality. Maybe not at eight, maybe not at twelve. But by fourteen? Why don't you ask them? It might be hard to pick a starting point, and you will probably struggle with the appropriate language, but unless they have a particularly open relationship with helpful family members, or access to a friend with that relationship, where are they going to find what they need to know about their sexuality?

8
Art

Can I see one?

"The arts are not just a nice thing to have or to do if there is free time or if one can afford it. Rather, paintings and poetry, music and fashion, design and dialogue, they all define who we are as a people and provide an account of our history for the next generation."

Michelle Obama

If we can't talk about the clitoris or read about it, can we see it? This chapter will consider visibility. It's a game of hide and seek—where is our protagonist to be seen when she's not living her private life, between the legs of most women? Just as there is a paucity of language when it comes to the clitoris, her artistic palette is similarly limited; if you don't want to look at what is considered porn, it's hard to see representations—and porn is deeply flawed. A 2017 study of PornHub's most-viewed videos found females were only portrayed orgasming 18.3% of the time, and on these occasions, this was signaled by facial expressions and vocalizations and did not appear to involve clitoral stimulation.[190]

Penises are visible everywhere in the Western cultural

world. They're as visible as breasts. They're in art galleries and on church ceilings, and public buildings and parks are littered with statuesque penises, peeing into fountains, flopping on the scrotum of beautiful marble Adonises, or living dangerously hanging around heroic men of action like Theseus and Perseus as they slay minotaurs.[191] Take five minutes for a Google Images search on Adonis statues or peeing statues, and you'll be amazed by their sheer numbers and familiarity. (And look out for Theseus on the Archibald Fountain in Sydney's Hyde Park—don't you think his weenie would have been safer in a jockstrap?)

We don't feel embarrassed when we see these statues or images, or flinch if we're with our grandmother, or mind when children point at them or comment. It becomes a small step from the flaccid penises of art to conversations with children at appropriate times and ages about what happens when they get big. They know what a penis is, and they know that generally men have them. The visual record lays the way to dialogue. But where do you start with the clitoris? If we're not *talking* about the clitoris, and there are no images of her, how on earth do we expect young women to learn about their anatomy? How does their site of sexual pleasure get endorsed? How will male partners know about the clitoris? Or ask questions about her? Or, importantly, treat her as an equal?

Vulva display

This wasn't always the way, as we've seen with prehistoric sculptures. There's more: lodged in the walls of some 11th- and 12th-century buildings in Europe are a surprising and joyous number of carvings of vulvas on display—the so-called Sheela-na-gig girls (no evidence of Australian slang in the etymology, apparently). The carvings portray women young and old, plump and skinny, with vulvas of various sizes and shapes. The women stand alone and look you squarely in the eye. A favorite of mine is the Milanese woman who stood defiantly in an archway above Porta Tosa—one of the eleven medieval entrances into Milan—which was renamed Porta Vittoria in the 19th century. That is, until she was removed to save her modesty or that of those who gazed on her. She now resides at the Sforza Castle Museum of Ancient Art in Milan.

Carving from the Porta Tosa, Milan, c. 1185

Her dress indicates that she is a woman of status and wealth. Her hair is carefully braided, so she's not disheveled or wanton. Around her neck is a dramatic jewelry cross, a symbol that she is Christian and therefore part of the societal mainstream; she is not a hag or harridan from mythology. Her gaze is direct. In her right hand is a large knife, held above her mons pubis. Her gesture is defiant, perhaps a warning to those entering the city through the gate. Academia cannot be sure of her mission—a legend says that in 1162, during one of Barbarossa's many sieges of the city, a young Milanese woman climbed the ramparts and, in an act designed to humiliate the enemy army, exposed herself. Another story suggests she was placed there as an insult, above the easternmost gate of the city and facing Constantinople, whose Empress

Leobissa had refused requests from a Milanese delegation for financial assistance. Either way it is hard to find shame or pornography in the carving, although it has been slightly spoiled for me by the suggestion she's about to shave off her pubic hair, or is holding a dildo rather than a knife. What was this vulva-displaying bas-relief doing standing above one of the main entrances to the city? Can you imagine it being commissioned today?

There are more than 100 medieval Sheela-na-gig or vulva-display carvings in Britain alone, many of them still on the walls of churches and castles. Some are positioned for all the world to see, while others are high up, speaking only to the elements. There used to be more, but over time, as society became increasingly anxious about representations of woman's sexual anatomy, they were removed or their lower bodies were chiseled or smashed off.

The medieval Sheela-na-gig carvings draw on a much earlier tradition. Plutarch tells a story about the bravery of the Persian women who raised their skirts to expose their naked vulvas, goading their own retreating army to re-enter the battle they were fleeing from. The Persian soldiers returned to the battle reinvigorated and emerged victorious.[192] The vulva display is presented as a gesture that brings shame on the retreating soldiers. Plutarch's story is the fifth in his series of anecdotes describing the bravery of women, and there is not a hint of any immodesty or of embarrassment about their actions. This moment is portrayed in the old master paintings *Bravery of the Persian Women* and *Persian Women,* by the Flemish painters Frans Francken the Younger and Otto van Veen respectively.

In Greek mythology, the invincible hero, Bellerophon—tamer of Pegasus the winged horse, slayer of the monstrous Chimera, conqueror of the Amazons, and friend of Poseidon, god of the sea—is thwarted in his attempt to take the city of Xanthos by a wall of women similarly exposing themselves. The sight of their vulvas forces back the waves, terrifies the flying horse, and shames the poor guy into retreat. The Irish sun god, Cuchulain, came a cropper in similar circumstances. Respect to the vulva! The Greeks even had a term for it, *anasyrma*, which literally translates as "skirt up." The more recent incidents of "upskirting," when boys and men have taken photos on their phones looking up unsuspecting women's skirts, illustrate neatly how respect for the power of the vulva has declined over time.

Did these legends and myths about the innate power of the vulva to subdue or shame men contribute to the desire to cover her up? Maybe the men who defaced the lower torsos of the Sheela-na-gig statues were worried about being another Bellerophon or Cuchulain. Perhaps the graphic reminder of where they came from was too humbling for these men to contemplate.

I struggled at university with the writings of Julia Kristeva, but the treatment of the Sheela-na-gigs seems a perfect visual of the attempt "to keep a being who speaks to his God separated from the fecund mother." She states that "[f]ear of the archaic mother turns out to be essentially fear of her generative power" and argues that it is this power that patrilineal societies seek to subdue.[193]

Or did the statues trigger pornographic thoughts? Thoughts of a sexual nature, not of the power of birth. After all, strong men like Samson have been undone by lust. Whatever its reason, the damage to these carvings sent men and women a message about the vulva: that it was to be despised, hidden, or smashed off.

While researching the Sheela-na-gig girls I came across a BBC piece about them, published in 2019, that contained a warning, in bold: "This article contains some graphic imagery." It referred to pictures of the bas-reliefs. Peeved, I searched for news stories with images of statues with penises, and soon found one from 2015 about an exhibition of Greek sculptures being curated at the British Museum. The final picture showed Ajax with the most enormous erection, but there was no warning about graphic imagery. In this day and age, please, BBC, can we have sex equality (now that you have sorted out salaries)?

Let's examine for a moment the purpose of the original medieval Sheela-na-gig carvings. Are they:

a) Remnants of a pagan Celtic belief system about a goddess, inserted into buildings by either recidivist stonemasons or canny clergy wishing to appease their flock?

b) Fertility totems?

c) For sex-education purposes?

d) To warn onlookers of the sins of the flesh[194] or to remind them of the lustfulness of women?

e) To ward off evil spirits?

f) "Hideous historical oddities"?[195]

The most popular theories have been that they are a), remnants of pagan times, which feels credible, or that they were d), lessons in lust and the nature of women, commissioned by clergy as warnings. We are back to the *querelle des femmes*. Remember though that these are only theories put forward by male commentators in the past; now, fortunately, academia is re-evaluating ideas about prehistoric and medieval vulva representations, although there is still an inclination to describe this renewed interest as having a feminist agenda. As if the male academics who argued that the Stone Age vulvas were early pornography didn't have an agenda. All academia has an agenda, defined by the society it operates in. It's hard to step away from one's culture, and to recognize how it affects one's lens in extraordinarily subtle ways. It affects the patterns one sees in a data set, the interpretations one puts on those data sets, and the types of data sets one seeks in the first place. Of course the clerics who destroyed or removed the Sheela-na-gig girls had agendas. The 19th- and 20th-century theories were misinformed and misguided products of their time, and now we can contextualize more appropriately for our time, through a lens that allows women and their vulvas to have played an inspirational role.

Ironically, historians have been keen to claim the Cerne Abbas giant, pictured here . . .

The Cerne Abbas giant, Dorset, England.

. . . as yes, you've guessed it, an ancient fertility symbol! What else could he be? Early pornography? Homoerotic art? A warning to young girls of the rapacious nature of men? A celebration of the greatness of clubs and penises, and the smallness of men's brains? (Well, just look at the size of his head.) Although no evidence dates the Cerne Abbas giant to earlier than the 17th century, he has been wishfully referred to in much literature as prehistoric. The stature of the penis is so embedded into our society that its dominance is a given, imbued with ancient solemnity. And BTW, is that a vulva in the background?

Visibility in the art world

Every so often, to highlight how few female artists are being exhibited, a group called the Guerrilla Girls counts the number of artworks by female artists versus the number of female nudes on exhibition in national galleries. The bright yellow Guerrilla Girls 2012 posters and postcards showing a reclining female nude wearing a gorilla mask are unforgettable. They have the headline, "Do women have to be naked to get into the Met Museum?" followed by the statistics, "Less that 4% of the artists in the Modern Art sections are women, but 76% of the nudes are female."

Inspired by their iconic works, I decided to count the number of naked sculptures and paintings of men versus women at the Art Institute of Chicago, one of the top art galleries in the world, making a note of the kinds of poses naked men and women were portrayed in. I knew that Renaissance, neoclassical, and Romantic works would throw up equal numbers of men and women, both in sculpture and paintings, but that the women from these eras would be in coy poses, while the men would be shown in more dynamic stances, fighting wild animals or in battle. Art has often reflected widely held beliefs about women and their role in society, possibly because so much has been generated by men: when a woman got a rare opportunity, she had to conform or risk losing her standing.

Like the Guerrilla Girls, I decided to focus on what the hopefully more enlightened Modern and Contemporary sections would show the world about the anatomies of men and women. In a twist on looking at the gender of the artists being displayed (mostly men) versus the nudity shown (mostly women), I wanted to see whether the clitoral invisibility found in language is also reflected in art, in contrast to the visibility of the male sexual anatomy. With so many female nudes around, it seems a relevant question. While anatomically our little clitoris is less visible, do the poses naked women are shown in still shield her, or her location, from display? Haven't advertisers been telling us for a while that women can wear a tampon *and* run and climb and hike and do all the things that men do?

I was hoping Lucian Freud would oblige; he is well known for his unfaltering fleshiness. But though the Art Institute of Chicago does own Freud's *Sunny Morning—Eight Legs,* depicting a naked man, a dog, and another pair of legs, it wasn't on display. Later, scouring through a catalog of his work in the institute shop, still eager to find a painting of our little clit by an acknowledged world great, I found that while there is no shortage of manspreading, Freud's women have their legs firmly closed or positioned side on. It's not that he didn't know, or want to know, what was going on between the legs of the women he painted: with 14 acknowledged children, perhaps a further 40 unconfirmed, and more female lovers than one can count, Lucian Freud was as close to being a "notorious heterosexual" as it's possible to get.[196] Nonetheless, liberal as he was, he didn't paint clitorises.

When I visited, the institute had a gallery full of gorgeous Georgia O'Keeffe paintings, notably her *Yellow Hickory Leaves with Daisy* from 1928. The commentary for the picture explained:

> In *Yellow Hickory Leaves with Daisy*, O'Keeffe painted
> the leaves around the small white daisy at lower center
> so that they seem to be emanating from the flower.
> She hoped that the strangely magnified subjects
> would inspire viewers to, as she said, "be surprised
> into taking time to look at" them in a new way.[197]

I, like other feminists, long for Georgia O'Keeffe's flower
pictures to be representations of vulvas, asking viewers to look
at them in a new way. However, O'Keeffe was always adamant
that this was not the case. She was painting flowers. Large,
glorious, and gorgeous flowers. On the one hand, I want to
respect what O'Keeffe says about her work, because it's annoy-
ing and rude when someone else tells you what you're thinking
when you've clearly stated a different point of view. This is
exactly what has happened to women and their sexuality, and
I'm angry about it. On the other hand, putting these instincts
aside, I can't help thinking that sometimes a flower is just a
flower, and sometimes it is also a vulva, with a clitoris, whatever anyone says. If you get the chance, I urge you to go and
see or search for images of her beautiful pictures online and
decide for yourself. For the purposes of this book, O'Keeffe's
paintings can't be recorded as representations of vulvas and
clitorises. That leaves the number of vulvas and clitorises on
show in the Modern section at Chicago's art institute as . . .
zero. Zippo. None. Not one.

   In another game I played at the institute, where the winner
was the person who spotted the most penises in an hour, my
son won, with 29. (He got lucky in the photography section.)
Naked women aplenty, sculpted and painted, with ankles,
hair, gorgeous breasts, and curvaceous bottoms on show, but
not clitorises. Rarely even a vulva and especially not one with
pubic hair. In 2011, when Frédéric Durand posted a picture of
Gustave Courbet's 1866 *The Origin of the World* on his Facebook

page with a link to an article about the painting, his account was deactivated for sharing pornography. Yes, we need to be clear about where we stand on pornography and exploitation, but if a picture of a woman's hairy mons pubis (not even splayed) is pornography, why aren't all those public penises pornographic? For the world of high art, it's as if the vulva and clitoris don't exist. Naked women are welcomed into galleries, as the Guerrilla Girls highlight, but the finer details of their anatomies are not.

Like Orpheus following Eurydice into the underworld, I could follow our clitoris into the murky underworld of "scurrilous" prints, photography, and porn, to see whether I could find her and bring her to the surface. However, even if I did find her it wouldn't solve the problem of her visibility for children or teenagers, or many adults for that matter, who choose not to access that kind of material. In pornography, the focus is always of a sexual nature. For many women, their early forays into pornography were met with fantasies directed by men and aimed at men, and women didn't stay long; what they saw didn't chime with their needs. My first pornography viewing, in the 1980s, showed me Anna, who was on her knees giving head *and* managing to gasp breathlessly between dives, "Umm . . . This is so goooood . . . I love your big cock . . . your really big cock . . . I want you to come in my mouth . . ." before her head was pushed forward and held in place by the recipient's hands. I was choking and gasping for air at the thought. I understand the erotic nature of a classic BJ for men; I didn't see myself reflected in Anna's enthusiasm. After a brief foray more recently, I still found Anna but now she is with Mark, who also spends time on his knees, doing a lot of one-sided licking to allow for the camera to capture the action. Anna is now murmuring, "That feels so . . . soooo good . . . I want you inside me," and I want to shout at the screen, "FFS, just finish what you started!" One reason women give for not having an orgasm during their sexual encounters is that they don't get enough clitoral stimulation, and foreplay too quickly moves on to coitus.

Even if you feel comfortable with pornography—its content and its uneasy relationship with sexual exploitation—it's still hard to navigate oneself to woman-centered erotica. My guess is that if you find it, you already know all you need to know about your clitoris. On my last trip to the U.K. I opened one of the mainstream Sunday magazines and found an article by columnist Hugo Rifkind, who'd visited a female-friendly-porn movie set. Rifkind describes the filming of a scene that opens on a bed where a man is mechanically having sex with a bored-to-death woman, when Super Clit appears, looks at them wearily, and flings the man across the room into an armchair, before berating Jane for being a sex toy and not knowing how to please herself. Great! But Rifkind also found that the most-watched videos on Pornhub (the largest porn site in the world) had titles that included the words "cute teen stepdaughter," "surprise anal sex," and "sugar daddy gropes his sugar baby and her best friend."[198] Most teens don't pay for their porn and take pot luck with what they can find for free. Peggy Orenstein quoted a study of behaviors in such popular porn, in which nearly 90 percent of the 304 random scenes contained physical aggression toward women, while close to half contained verbal humiliation.[199] There is little reverence for either the vulva or female desire and pleasure being modeled here.

Surely the place for young women and the uninitiated to find out about sexual pleasure and to see representations of the clitoris is in sex-education and anatomy books. Sadly, we know what happens there. The clitoris is invariably passed over, barely represented as a small dot, and overshadowed by the miraculous childbearing capacity of a woman's body.

Why do we remain so uncomfortable with the vulva and clitoris? Outside the world of pornography, there is an aversion to seeing even clothed women with their legs splayed. A womanspreading placard made in 2019 by Rumisa Lakhani and Rashida Muqadam for an International Women's Day march in Pakistan encapsulates the taboo perfectly. Rumisa says of her image—which depicts a chic woman modestly clothed in pants, a long-sleeved top, and a scarf sitting with her knees apart and her hands on her thighs—that "women have to be elegant; we have to worry about not showing the shape of our bodies. The men, they manspread and no one bats an eye." Rashida's slogan translates as "Here, I'm sitting correctly." Their placard caused outrage.[200]

Rumisa and Rashida's womanspreading placard.

We are no more comfortable with women sitting this way in the West. Men are not discouraged from sitting with their legs splayed—they do it all the time, in business meetings, on park benches, on the bus. There is no question of their crown jewels being more discreetly attired or aired. There is not a hint of provocation implied when a man sits like this; he's just relaxed. Yet when a woman positions herself similarly, it's deemed unladylike and risks being read as a come-on. Men are allowed to flop their hands into their crotches without it being seen as a provocative hip-hop dance move, and they can gently readjust their bits to achieve greater comfort without it being any more shocking than hitching up a sock or repositioning a lock of hair. They can display their clothed crotch area to all and sundry without censure. We've come a long way from the days when women were not allowed to show their ankles and had to wear long skirts or dresses, but there is still a clothed modesty about the area between a woman's legs. It's as if by splaying them she opens the way to her vulva—and this is taboo. I care about this taboo, because by censuring the vulva we limit knowledge. We imply there's something to be ashamed of and what's there is not decent—or why would it need to be kept hidden? Come back, women of Persia, we need you!

In 2013, the fabulous, trailblazing mixed-media artist Sophia Wallace started the Cliteracy project. In an interview with Abby Martin for the RT news channel, Wallace described her viral project as "a movement, a call to the world to finally recognize female bodies and female sexuality are on an equal par with male bodies." She started by exhibiting her news-print posters, creating a 10-by-13-foot-high wall of them called *Cliteracy, 100 Natural Laws,* and pasting them individually onto the streets of Brooklyn like old posters. They said things like "The hole is not the whole" and "Freedom in society can be measured by the distribution of orgasms." Wallace and Center Artist Labs rented a huge Santa Fe billboard next to a highway and plastered it with the words "Democracy Without Cliteracy Phallusy." Wallace wanted to set the clit large, and in public spaces. In 2013, she ran a project with artist Ken Thomas called the Clit Rodeo, in which a giant, gold, rideable clit was mounted on a spring, like a little red tractor or wooden horse in a children's playground. People were invited to ride the clit, respect the clit, and have fun. They did! In 2014, the Whitney in New York staged an art intervention, the Clitney Perennial, and Wallace was invited to give a TED Talk. In terms of art, it's contemporary mainstream, which is great news for the clit, and I'm excited for Wallace's next project. When I started this book, a close friend gave me a gold pendant purchased from Wallace's website as a totem, and a photo of it became the image for my Instagram bio, @its.personalgirls. Inspired by Wallace, the clit sometimes makes its spray-painted mark as graffiti. I follow people on Instagram who sculpt, paint, knit, or sew vulvas and clitorises into gorgeously colored art, sometimes selling it on Pinterest. My daughter put the clit into a rug design that she painted onto our deck. But most people still don't recognize it for what it is. Laura Dodsworth's project *Womanhood* (2019) exposes and reclaims the vulva through 100 photographs and stories from vulva owners and received mainstream media coverage. The taboo is being eroded.

Open any well-used high-school textbook and you will find the same crude drawing of a penis replicated, like an emoji, over and over again, drawn with blunt pencils in the margins of *Macbeth, Of Mice and Men,* and *As You Like It.* At one school, I sometimes taught in a classroom where there was a penis in indelible ink on the back of every chair. I used to apologize to the girls in my sophomore class, saying I was sorry they had to look at that part of a man's anatomy uninvited for 40 minutes, and we would laugh. This was just part and parcel of the learning environment. Students, from the age of 11 upward, were exposed to a daily barrage of erect penis images, normalizing the sight of the male member through such crude but generally accurate cartoons and therefore endorsing and to a certain extent glorifying male sexuality. I don't think it has a place in a classroom unless it is a sex-education lesson, but, I wonder, why is the graffiti penis always erect? Is it because it's more impressive that way? More powerful? Don't get me wrong, I'm not a prude, but I am for equality and I didn't once see an image of a vulva. If I had, would there have been a clitoris drawn in? Let alone a swollen, aroused one?

I raised this once with some of my more robust senior students. Part of the school-leavers' graduation ritual was on the final day of school, when students brought in indelible marker pens and asked friends and teachers to sign their shirts. Inevitably every boy in my set 4 GCSE English class, the class of Darren, Dan, and George, had penises emblazoned across their shirts by 10 a.m.

"'Ere, Miss, will you sign my shirt?"

"Yes, Miss, sign our shirts!"

"You can't tell us off for drawing on them today, Miss, we're allowed! Mr. Haydon [the principal] even signed!" They were gleeful and proud, bestowing compliments on me as a teacher by asking me to be immortalized on their shirts.

"Yes, but I won't sign if there's a penis on that bit of the shirt. Unless," I added in a calculated moment of feminist rebellion, "unless I can draw a vagina as well." (I wasn't sure they'd know the word *vulva* and I wasn't up for an anatomy lesson, and *cunt*, as discussed earlier, was out of the question. They'd have been shouting it down the corridors all day.) You should have heard the uproar.

"You can't do that!"

"That's rude, Miss."

One boy even told me confidently, "It's not allowed, Miss." That was George; you think he'd have learned his lesson.

I didn't draw a vulva, settling instead on signing next to

the smallest penis I could find, saying, "At least this one is to scale." I knew it was a cheap joke, but they liked it.

"You've got form, Miss."

"Yeah, that's right, proper banter that is."

God love 'em.

What would have happened if I'd emblazoned vulvas across those 16-year-olds' shirts? I really wish I had. Would it have been "allowed"? If I ever teach again, and catch anyone drawing penises, I'll haul them up for sexism and insist they draw a vulva, including a clitoris, next to every penis in the book, and in a few books from the storage closet as well, just to make up for the imbalance. I'll start with this visibility and then maybe we'll see more of her in the art world, eventually.

Do women want clits on display? I don't know, but I do know no one has ever asked them, and society has set women up to feel that the clitoris is taboo and an inappropriate part of the anatomy to show in public. There was a time in the 18th century when a woman's ankle was shocking. Men don't seem to have a problem with penises being so visible in culture. The point is, if a girl can't read about her clitoris, or see it in the world around her, how's a girl to know she's got one?

I had been wondering for a while what a vulva emoji might look like, then I came across www.flirtmoji.co, where you can download them along with a fine array of other sexy text aids. The vulva's representation in emoji form will give her cartoon visibility, which is a good, nonthreatening place to start raising her profile. However, we should ensure that emojis are accurate. Any sexting with a clitoris-free vulva should be bounced back, swiped left, responded to with a flaccid penis emoji, or just plain ignored. There is also, of course, the issue that the nub of the clitoris is just the tip of the iceberg—and I haven't yet been able to find a clit emoji to download to my phone.

The pictorial maxim of the three monkeys depicting Confucius's code of conduct sums up attitudes to the clitoris both in the past and in society today.

See no evil   Hear no evil   Speak no evil

PART FOUR:
CLIMAX

9
Yes, Yes, Yes to Cliteracy!

"Women's Studies can amount simply to compensatory history; too often they fail to challenge the intellectual and political structures that must be challenged if women as a group are ever to come into collective, nonexclusionary freedom."

Adrienne Rich, poet and essayist

"Too much of a good thing can be wonderful!"

Mae West, entertainment icon

My fabulous editor, who, like me, grew up in the U.K. but now lives in the States, shared an anecdote with me.

"How do you say *clitoris* here? Is it clit-*or*-is or *cli*-toris?" she asked a friend.

Her friend, Susan, laughed and replied, "We don't actually say it."

This reminds me of the *Seinfeld* episode when Jerry shouts out "Dolores!" (to rhyme with "clit-*or*-is"). When I was pitching this book to agents, I had a couple tell me that they liked my writing but the subject was "niche," as if it was some BDSM kink. When my mother told her friends about my book, most were enthusiastic, but a few said things like, "That's a shame. Does she have to?" Yes, I have to! The history of the clitoris is bound up with the history of women. It has been repressed. It is one of the last taboos, along with intersex bodies, but it's only a matter of time before the Western world recalibrates. Capacity for and engagement in sexual pleasure is not a measure of morality in anyone. All people, however they identify, sometimes behave in immoral ways when it comes to sex, just as they do in other areas of life, like money. Research in 1991 found that women who enjoy sexual self-pleasure "had significantly more orgasms, greater sexual desire, higher self-esteem, and greater marital and sexual satisfaction"[201] and we know women who are comfortable using the word *clitoris* are more likely to orgasm as part of partnered sex. A sense of personal sexual wholeness is incredibly important to emotional well-being, and being allowed to own one's sexual truth, without shame, is fundamentally empowering. It is time to give vulva owners their clitorises, and to celebrate their wonder.

What is the full nature of female sexuality and desire? The truth is we don't know, because it has been repressed and inaccurately constructed by faulty and misguided science, religion, philosophers, psychologists, and policymakers. Do vulva owners like sex less than people with a penis? Is emotion and tenderness more important to them than sexual satisfaction? Is there a problem with the orgasmocentric construct of sexual satisfaction? Let vulva owners decide. Give them the knowledge and space to know for themselves.

Much research about female sexuality and its implications has been skewed by biases forced on women by history. After looking at research on gender differences in attitudes to sex, I've found that differences are often rooted in the stigma against women expressing sexual desire, women's socialization to attend to others' needs rather than their own, and the double standard that dictates different sets of appropriate sexual behaviors for men and women. We should take another look at some of those (mostly heteronormative) timeworn studies. Is it really true that men think about sex more than women think about sex?[202] Roy Baumeister's famous 2001 study—which became legendary proof that men thought about sex more than women—was replicated by Terri Fisher, Zachary Moore, and Mary-Jo Pittenger in 2011 to measure the number of times male and female participants thought about food and sleep (as well as sex) over the course of a week. The result? "Yes, men thought about sex modestly more frequently than women did. However, men also thought about both food and sleep significantly more often than women did. Thus, men reported a greater number of personal-need-based thoughts than did women overall."[203] This study ends up telling us more about the way men and women dwell on their own needs—or have been encouraged or allowed to dwell on them—than it does about their essential sexual natures.

The fact that men ponder their own physical needs more could go a long way toward explaining why casual hookups with cis men return such a low orgasm rate for women.[204]

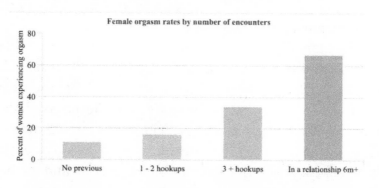

Men are more comfortable placing an emphasis on their own needs; society encourages women to be facilitators and hasn't encouraged them to demand satisfaction (in such things as promotions or orgasms). It's also true that vulva owners generally don't figure out what they like, or how to orgasm, until later than penis owners, through lack of information and the stigma that still surrounds female masturbation. In terms of sex education, we haven't moved on from Marie Stopes's observation over 100 years ago that there is a universal lack of knowledge. I was shocked last week to overhear a sassy 15-year-old girl whose parents I vaguely know telling a friend about her latest escapade with a boy (he'd had much pleasure, she very little) and explaining that girls don't come anyway. If she's old enough to be going down on him, and he has access to her vulva and vagina, I'd at least like her to know her options. And how are young cis men to know about vulva owners' needs if we're not telling them and they're reliant on porn?

The oft-repeated statistic that 70 percent of men will accept a casual invitation to have intercourse while zero percent of women will (leading to claims that men like sex more), from the Russell Clark and Elaine Hatfield research referenced earlier, changes dramatically if that casual sexual offering to a woman is perceived to be one that will result in great sex.[205] Across multiple studies, the perceived sexual capability of the proposer most strongly predicts the acceptance of casual-sex offers among both men and women. It seems the initial experiment told us more about women's actual experiences in casual hookups than it did about women's desire for them. Julio González-Alvarez and Teresa Cervera-Crespo concluded in their research that men disassociate sexual attraction from moral judgment more than women,[206] but is this an essential aspect of being either a man or a woman, or is it a finding that reflects sexual socialization? Historically the evolutionary narrative has been that women seek breeding and home-providing mates, while men seek multiple mates, and that for women, sex and morality are symbiotic—but is this the whole sex story? How does contraception change the landscape for women who have most at stake with an unwanted pregnancy? Are vulva owners not able to discern between a passing fancy, a medium-term hook up, and a long-term mate? Might they function differently if the script was less proscriptive? If the sex was better? Safer?

This book is about the clitoris and argues strongly that female sexual pleasure and the role of the clitoris should be normalized. However, female sexuality is far more complicated than the clitoris. It *is* a primary site for orgasm, but we are only just understanding the role of the brain in orgasm. Recent research has observed women orgasming with stimulation "from many body regions as well as imagery alone. Orgasm is not just a reflex, it is a total body experience."[207] Hopefully these findings will be met with enthusiasm rather than denial. There is still so much to learn about how sexual pleasure can be experienced if we are open to the possibilities.

There are many reasons for vulva owners to enjoy intercourse, among them arousal, enjoyment of physical intimacy, and the expression of affection it can represent for both partners. And for some, orgasm. I am not making an argument against intercourse; I am making an argument for orgasm to be part of the sexual experience that surrounds intercourse so that vulva owners aren't shortchanged. Society needs to carefully distinguish between sex for pleasure and reproductive coitus. Most sex that occurs today is for the former and not the latter purpose. The model of reproductive intercourse as *the* sex act marginalizes the clitoris because, within this model, it doesn't have an essential role.

Reproductive sex has changed greatly over the past 50 years. With increasing medical knowledge and the ability to monitor ovulation and freeze eggs, the conception of a baby can become as much about planning and timing as it is about having intercourse. In the United States, many people visit their gynecologists prior to beginning intercourse for conception. Assisted pregnancies are more common. Couples hoping to conceive are advised to monitor intercourse frequency. For couples who are not conceiving as quickly as they would like, intercourse can become a task to fulfil at the optimum time to maximize conception. It can seem like hard work. The desire is for a baby, not for sex. The sexual element of orgasm is necessary for the penis, but much of the anticipation and pleasure is stripped away by the pressure to perform and the tension surrounding the unspoken question: will it work this time? If the model for sex shifted so that there were two strands—sex for reproduction and sex for pleasure—then the second could incorporate, but not be exclusively *about,* intercourse. It would be a more inclusive model of sexuality, and our clitoris would fare much better. The clit would have a definite role: provider of orgasms. Then the only remaining question would be why penis owners should have more orgasms than vulva owners in their sexual encounters. Let's separate out reproductive sex and create a more inclusive definition of sex that can involve—but is not exclusively about—P-in-V intercourse. This would enable us to educate *all* people about sexual health and sexual pleasure, which would surely benefit everyone's well-being.

The discourse around female sexuality is awkward and limited, but women today have a few advantages that we didn't have historically.

> We understand the anatomy of the clitoris in all its glorious completeness. Not only has the clitoris been fully mapped and dissected in cadavers, but it has been 3D imaged live. There is no need for confusion about how it functions.

We have access to data about how sexual pleasure works for cis women, and this proves that the clitoris is central to female sexuality. We should educate everyone—girls, boys, transgender, and intersex young people—*equally*, on all fronts, so that they can make informed choices about their sexual interactions and what their sexuality means.

Women and other vulva owners have more of a voice than they have had in the past. I urge them to use those voices and I urge others to listen. Use them! You're only asking for an orgasm, after all—not a pay rise.

We have many allies who are feminists and cheer-leaders: they will advocate with us.

Given this set of circumstances, now is the time to rewrite sex equality. Normalize sex education to include accurate information about pleasure for all participants, and normalize communication between partners to encompass pleasure as well as consent.

This book started with one malapropism, and ends with another. It comes from a girlfriend who recounted her efforts to make conversation with an American ambassador at a rather grand English summer drinks party. She gestured to the rampant flowering vine adorning the trellised garden wall and said, "Look at that glorious clitoris!"

May all your clematis be as glorious.

Appendix
FGM Facts

The many terms for this practice—female genital mutilation, female genital cutting, female circumcision, washing your hands, initiation into womanhood, washing oneself clean, ritual purity, purification[208]—are indicative of the complexity of discussing and dismantling it. In 1997, the World Health Organization, along with the United Nations Children's Fund and the United Nations Population Fund, issued a joint statement condemning the practice of FGM. Since then, these institutions have been working at international, national, and sub-national levels to eliminate the practice. The UN made eradicating FGM a sustainable development goal in 2012. The WHO currently estimates that three million girls globally are at risk of undergoing female genital cutting every year, joining 200 million other girls and women alive today who have already experienced it. The majority of girls are cut before they are 15 years old. While the procedure varies, most cases include the surgical removal, reduction, or partial removal of the clitoris. The facts that follow are by no means comprehensive. I give them as evidence of the range of FGM in terms of procedure, incidence, geography, and demographics, to highlight the issues that exist for anyone designing programs that will affect change.

Types of FGM

Female genital mutilation is classified into four major types.[209] It is usually performed by a traditional practitioner.

Type 1: The partial or total removal of the clitoral glans.

Type 2: The partial or total removal of the clitoral glans and the labia minora, with or without removal of the labia majora.

Type 3: Also known as infibulation. This is the narrowing of the vaginal opening through the creation of a covering seal. The seal is formed by cutting and repositioning the labia minora, or the labia majora, sometimes through stitching, with or without removal of the clitoral prepuce/clitoral hood and glans.

Type 4: This includes all other harmful procedures to the female genitalia for non-medical purposes, e.g. pricking, piercing, incising, scraping, and cauterizing the genital area.

Incidence of FGM
America and Europe
Before we criticize other governments for their failure to address FGM, we should consider the records of our own governments. A Centers for Disease Control and Prevention study published in 2016 estimated that half a million women and girls in the United States are at risk of or have been subjected to FGM. It notes the difficulty of obtaining accurate data. In November 2018, in a federal case, *U.S. v. Nargarwala,* a U.S. licensed doctor was charged with performing FGM on nine girls, aged seven to thirteen, at a Detroit clinic. The judge dismissed six of the eight charges on the grounds that the law was unconstitutional because Congress did not have the right to criminalize the practice. In early 2019, the Department of Justice decided not to appeal this ruling. The House of Representatives has now intervened, hoping for a clarification.

FGM has been illegal in the U.K. since 1985, but 2019 was the first time there was a conviction. A National Health Service survey, tracking women and girls using an NHS service where FGM was relevant to their attendance, found that 285 of them had had FGM undertaken within the U.K. between 2015 and 2020. Overall, 24,420 women and girls who had undergone FGM were seen.[210]

While FGM may be illegal, people are accessing it within their communities in Europe and America. Some travel to countries of family origin for "vacation cutting." Many activists question the West's level of commitment to enforcing change on its own soil.

International

A 2020 UNFPA–UNICEF joint program on FGM published data on the percentage of girls and women, aged 15 to 49, who have undergone some form of FGM.[211] It is drawn from the 17 countries in which data has been systematically collected.

Somalia	98%
Guinea	97%
Djibouti	93%
Egypt	91%
Mali	91%
Eritrea	89%
Sudan	88%
Burkina Faso	76%
Gambia	75%
Ethiopia	74%
Mauritania	69%
Guinea-Bissau	45%
Senegal	26%
Nigeria	25%
Kenya	21%
Yemen	19%
Uganda	1%

Demographics and FGM within countries[212]

UNICEF reports that in countries where the practice is common, 63% of men and 67% of women want it to end.[213] However, this statistic masks great variation. For example:

- In Kenya, there is a sharp contrast in the prevalence of FGM between women without education and women with the highest level of education (58% and 9%, respectively).

- In Egypt, more than 90% of women in rural areas have undergone FGM, as opposed to 77% of women in urban areas.

- In Mauritania, more than 90% of women from the poorest households have had any form of FGM, compared to 37% of women from the richest households. However, in Mali, women from the richest households have higher FGM prevalence than women from the poorest households (87% and 64%, respectively).

- In some countries, for example Burkina Faso and Guinea, it has been found that some religious communities are more resistant to change when it comes to FGM than others. No change in prevalence was observed among Muslim women in the past decade, while Christian, Catholic, animist, and non-religious groups show a decline.[214]

- Sudan has a high prevalence of FGM both in rural and urban areas (87% and 86%), but less than half of women in both areas support the practice—especially in urban areas, where only 28% of women think FGM should continue.

Reduction so far

UNICEF data shows an overall decline in the incidence of FGM in those countries where it is practiced most, from 51% of girls aged 15 to 19 having undergone FGM in 1985 to 37% in 2016. In Kenya, over three generations, the practice has almost disappeared among some ethnic groups, notably the Meru, Kikuyu, Kamba, and Kalenjin. Many of the African countries where FGM is endemic have passed legislation criminalizing FGM, although Reuters reports that half of all girls who have undergone FGM or are at risk live in three countries—Egypt, Ethiopia, and Nigeria—all of which have laws against FGM.[215]

Fuambai Ahmadu, an American academic who returned to her parents' country of origin, Sierra Leone, in December 1991 to undergo female initiation, explains that within the Kono community of Sierra Leone the procedure is an initiation ritual, making those that undergo it into the "kind of person that is admirable: informed, courageous, capable of dealing with pain, mature and womanly."[216] Ahmadu argues that the West stigmatizes those who have experienced female circumcision and creates a myth of sexual dysfunction, highlighting the strong emphasis within Kono culture on female sexual pleasure and the importance of female sexuality. I give this story to illustrate that for many women who undergo FGM, it is with pride and a sense of purpose. Ruth Goldstein, who lived with an extended Malian family as part of her work with UNDP, USAID, and UNICEF to understand FGM in Mali, talks about the difficulty of intersecting her opinion of FGM into a community where it was so highly regarded. "I could not figure out how to make my answer balance how I felt with how she felt."[217] And this is the problem of being an outsider.

Last autumn, I was invited to a talk by Chicagoan Amy Maglio, founder and director of Women's Global Education

Project. WGEP is an international NGO focusing on girls' education. The organization subverts the traditional top-down, big development approach by operating on micro-levels within communities to keep girls in school. By partnering with local community groups, WGEP supports girls, providing some scholarship funding, and also enables community workshops focusing on sex education, reproductive health, women's empowerment, and community leadership. One such initiative, in the Tharaka region of Kenya, has led to the introduction of an alternative rite of passage for girls approaching puberty that celebrates the girls' coming of age, without FGM. This success comes from a commitment to understanding the needs of the community, the cultural role of FGM within it, and through enabling the community to drive and embrace change. Donate if you can. **https://womensglobal.org**

Acknowledgments

D ear Laura Dail, my literary agent, without you there would be no book. Thank you for your spontaneous enthusiasm, insight, wisdom, and guidance. You were my wish come true. Maryann Karinch, you were my second wish. Thank you for believing in me as a writer, for your constant availability, and your professionalism. Truly, I got lucky.

I was also blessed to have a good friend and editor in Sophie Miodownik. Sophie gave me excellent advice on early drafts of this book, and it felt like a ringing endorsement when she agreed to work on *The Sweetness of Venus* for publication. I love the way she made my writing better.

I am fortunate to have grown up with a family—Mum, Dad, Kate, and Robert—who will all reliably line up behind me and any project that I suggest with two phrases: "You will be amazing at this," and "What can we do to help?" This is a huge gift, and I love you.

Emily, remember that coffee in Floriole when you said, "You should write a book on this"? Christine, when I told you about the book in hushed tones in the dining room at the Arts Club and you went quiet and then said, "I love it"? Lisa, when I told you—and the next week you gave me a Sophia Wallace clit as a totem? Catherine and Griff, when you were prepared to tell your literary friends about me? Your votes of confidence were vital.

I am hugely indebted to the University of Loyola and its generous library policy for guests. This is in the true spirit of Jesuit learning; without access to your resources, I literally could not have written this book. Thank you. My thanks also go to Jane and Anne, who were instrumental in helping me.

My friend and women's empowerment coach, Hélène

Stelian, thank you for your generosity and rigor when it came to focusing on my proposal. You pushed me, and it was a consult that made the difference. I am also grateful to Dr. Beverly Whipple, doyenne and trailblazer of research into female sexuality, for lending her encouragement to a lay writer.

I had many early readers: Sue, Briony, the mother-daughter duo Suzanne and Ana, Tash, Tania, Amelie, Megan, Alice, Eugenie, Robin Gayle at UIC, Rosie, and Stef. It was your honesty and conviction that the world needed this book that kept me going. If you want cheerleaders, you could not do better than Greg and Greg, Phil and Michael, Lily, Jen, Meghan, Robbie, Ben, Luke, and everyone at Laura Dail Literary Agency, especially Samantha Fabien who worked on the translation rights and Carrie Pestritto for her work on securing an audiobook. Bob Podrasky, I'm excited for this. All of you—your pom-poms are glorious.

Eilidh Doig, I love the cover you created, and the Instagram graphics you did for me. Emily Fritz, your typesetting is gorgeous. And thank you, Ryan and Amelie, for being my social media gurus. I had no idea that I would enjoy that journey. @its.personalgirls has been so much fun, and in my followers I have found an incredibly supportive and enthusiastic community. Kim Schiefelbein and Michael Maniaci, you kept me together body and soul.

Finally, my family. Tom, thank you for supporting me through this project with food, discussion, and a conviction that I could do it. To my children, Zach, Theo, Hermione, and Rawdie, you make me wake up happy every day, fill me with love, and open up my world. Don't stop bringing me treasures.

Notes

Abbreviations for some reference works:

Laqueur, *Making Sex*: Laqueur, Thomas, *Making Sex: Body and Gender from the Greeks to Freud,* Harvard University Press, Cambridge, MA, and London, 1990.

Lloyd, *Female Orgasm*: Lloyd, Elisabeth A., *The Case of the Female Orgasm: Bias in the Science of Evolution,* Harvard University Press, Cambridge, MA, 2005.

Havelock Ellis, *Psychology of Sex*: Havelock Ellis, Henry, *Studies in the Psychology of Sex,* F.A. Davis, Philadelphia, 1913 and 1927 editions.

Orenstein, *Girls and Sex*: Orenstein, Peggy, *Girls and Sex: Navigating the Complicated New Landscape,* Harper Collins, New York, 2016.

Part One: Anatomy

1   von Staden, Heinrich, *Herophilus: The Art of Medicine in Early Alexandria,* Cambridge University Press, 1989 (quoted in Laqueur, *Making Sex,* p. 4). None of Herophilus's works survive, but he lives on in the writing of others, for example, Celsus, c. 25 BCE–c. 50 CE; Tertullian, 155 BCE–c. ?240 CE, and Galen, 129 BCE–c. 210 CE.

2   Criado-Perez, Caroline, *Invisible Women: Exposing Data Bias in a World Designed by Men,* Abrams Press, New York, 2019.

3   www.theguardian.com/lifeandstyle/2019/feb/23/truth-world-built-for-men-car-crashes

4   Aristotle, *Politics,* Book 1, Section 1254b, 350 BCE.

5   Gilbert, S. F., *Developmental Biology,* 6th Ed., Sinauer Associates, Sunderland, MA, 2000, Ch. 17, "Sex Determination" (quoting a translation of Galen, c. 200 CE).

6   Phrase coined by Thomas Laqueur in *Making Sex.*

7   Gallagher, Catherine and Thomas Laqueur (eds.), *The Making of the Modern Body: Sexuality and Society in the Nineteenth Century,*

University of California Press, Berkeley, CA, 1987, quoting Galen.

8   Kinsey, Alfred, *Sexual Behavior in the Human Female,* W. B. Saunders Company, Philadelphia and London, 1953, p. 580.

9   Gravelle, Karen, *What's Going On Down There? A Boy's Guide to Growing Up,* Bloomsbury Children's Books, New York, 2017, p. 59.

10   For example, *De naturalibus facultatibus* ("On the natural faculties"), 1523, tr. Thomas Linacre. However, there are many texts and translations of Galen. For a full list, see Richard J. Durling's article, "A Chronological Census of Renaissance Editions and Translations of Galen," *Journal of the Warburg and Courtauld Institutes,* Vol. 24, No. 3/4.

11   Vesalius, Andreas, *De Humani Corporis Fabrica Libri Septem,* Book 5, *The Organs of Nutrition and Generation,* 1543.

12   Margócsy, Dániel; Somos, Mark; Joffe, Stephen N., "Sex, Religion and a Towering Treatise on Anatomy," *Nature,* No. 560 (7718), August 2018, pp. 304–305.

13   Laqueur, *Making Sex,* pp. 86–91.

14   "Some creatures develop in such a way that they have two generative or one male, the other female. Always, when this redundancy happens, one of the two is operative and the other inoperative, since the latter, being contrary to nature, always gets stunted so far as nourishment is concerned; however, it is attached, just as growths (or tumors) are." Aristotle, *De generatione animalium* ("Generation of animals"), tr. A. L. Peck, Harvard University Press, Cambridge, MA, 1943, 772b26-32. www.archive.org

15   Lanfranco da Milano, *Lanfranci maioris,* in *Cyrurgia Guidonis de Cauliaco: et Cyrurgia Bruni, Theodorici, Rogerij, Rolandij, Bertapalie, Lanfranci,* Venice, 1498 (quoted in Leah DeVun, "Erecting Sex: Hermaphrodites and the Medieval Science of Surgery," in *Osiris,* Vol. 30, No. 1, p. 22).

16   Ibid.

17   Ibid.

18   Arboleda, V. A.; D. E. Sandberg; E. Vilain, "DSDs: Genetics, Underlying Pathologies and Psychosexual Differentiation," Nature Reviews Endocrinology, No. 10, 2014, pp. 603–615.

19 Soh, Debra, *The End of Gender: Debunking the Myths About Sex and Identity in Our Society,* Threshold Editions, New York, 2020.

20 Moreno, Angela, "In Amerika They Call Us Hermaphrodites," www.isna.org/books/chrysalis/moreno

21 "The Most strange and admirable discoverie of the three witches of Warboys, arraigned, conducted and executed at the last Assises at Huntingdon, for the bewitching of the five daughters of Robert Throckmorton Esquire, and divers other persons, with sundry Devellish and grievous torments: And also for the bewitching to death of Lady Crumwell, and like hath not been heard of in this age." Produced by the judge, Edward Fenner, and in collaboration with some others. Widdowe Orwin, for Thomas Mann and John Winnington, London, 1593. www.archive.org

22 Tomlinson, R. G., *Witchcraft Trials of Connecticut: The First Comprehensive, Documented History of Witchcraft Trials in Colonial Connecticut,* The Bond Press, Hartford, CT, 1978.

23 Parent du Châtelet, Alexandre, *De la prostitution dans la ville de Paris, considérée sous le rapport de l'hygiène publique, de la morale et de l'administration,* J. B. Baillière et fils, Paris, 1857.

24 Dapper, Olfert, *Descriptions of Africa,* Jacob van Meurs, Amsterdam, 1668, London, 1670 (quoted by Camille Nurka and Bethany Jones, "Labiaplasty, Race and the Colonial Imagination," *Australian Feminist Studies,* Vol. 28, Issue 78, December 2013, pp. 417–422).

25 Ten Rhyne, Willem, *An Account of the Cape of Good Hope and the Hottentots, the Natives of that country,* 1686 (quoted by Thomas DiPiero, *White Men Aren't,* Duke University Press, Durham and London, 2002, p. 126).

26 Bland, Lucy, "Trial by Sexology? Maud Allan, Salome and the Cult of the Clitoris Case," in *Sexology in Culture: Labelling Bodies and Desires,* ed. Lucy Bland and Laura Doan, University of Chicago Press, 1998, p. 188.

27 Fortescue, Captain the Hon. Seymour John, R.N., *Looking Back,* Longmans, Green and Co., London, 1920, p. 188.

28 Grant, Thomas, *Court Number One: The Old Bailey Trials that Defined Modern Britain,* John Murray, London, 2019.

29   "Hysteria," Merriam-Webster.com Dictionary, Merriam-
      Webster. Accessed October 13, 2020.

30   Gilman, Sander L.; Helen King; Roy Porter; G.S. Rousseau;
      and Elaine Showalter, *Hysteria Beyond Freud,* University of
      California Press, Berkeley, 1993.

31   Maines, Rachel P., *The Technology of Orgasm: "Hysteria," the
      Vibrator, and Women's Sexual Satisfaction,* Johns Hopkins
      University Press, Baltimore, MD, 2001.

32   Pieter van Foreest, tr. Schleiner, 1995, p. 113 (quoted in Helen
      King, *Galen and the Widow: Towards a History of Therapeutic
      Masturbation in Ancient Gynaecology, Journal on Gender Studies in
      Antiquity,* No. 1, 2011, pp. 205–235).

33   Maines, *Technology of Orgasm,* p. 10.

34   https://medical-dictionary.thefreedictionary.com/
      hysterical+paroxysm

35   Maines, *Technology of Orgasm,* p. 2 and notes, p. 126.

36   Hollick, Dr. Frederick, *The Diseases of Women; Their Causes and
      Cure Familiarly Explained with Practical Hints for Their Prevention
      for Female Health,* T. W. Strong, New York, and Cottrell & Co,
      New York, 1847, pp. 206, 210, and 211.

37   Monell, Samuel Howard, *A System of Instruction in X-Ray
      Methods and Medical Uses of Light, Hot-Air, Vibration and High
      Frequency Currents,* E. R. Pelton, New York, 1902, pp. 591, 606,
      and 613.

38   Gully, James Manby, M.D., *The Water-Cure in Chronic Diseases:
      An Exposition,* Fowler and Wells, New York, 1870, p. 185.

39   Pearce, John M. S., "Sydenham on Hysteria," *European Neurology,*
      No. 76, 2016, pp. 175–181 (quoting from Sydenham's complete
      works, tr. John Peachey, M.D., London, 1742).

40   Trall, Russell Thacher, *The Health and Diseases of Women,* Office
      of the Health Reformer, Battle Creek, MI, 1873, pp. 7–8.

41   Maines, *Technology of Orgasm,* Ch. 4.

42   Hollick, *Diseases of Women,* p. 214.

43   Thompson, Lana, *The Wandering Womb: A Cultural History of
      Outrageous Beliefs About Women,* Prometheus Books, Amherst,
      NY, 1999, p. 69, quoting Zacuto (1452–1515), see also Moxius
      (1587–1612), p. 70.

44 Pope, Curran, *Practical Hydrotherapy: A Manual for Students and Practitioners*, Lancet-Clinic, Cincinnati, 1909, pp. 181, 510–12.

45 Griesinger, Wilhelm, *Mental Pathology and Therapeutics*, New Sydenham Society, London, 1867, reprinted Hafner, New York, 1965, pp. 179–81.

46 Carter, Robert Brudenell, *On the Pathology and Treatment of Hysteria*, John Churchill, London, 1853, p. 69.

47 www.livescience.com/sexist-medical-ideas-about-women.html

48 Showalter, Elaine, *The Female Malady: Women, Madness, and English Culture, 1830–1980*, Virago, London, 1987.

49 Venette, Nicolas, 1696, pp. 20–21 (quoted in David F. Greenberg, *The Construction of Homosexuality*, Chicago University Press, 2008, p. 374).

50 Estienne, Charles, *De Dissectione Partium Corporis Humani*, Libri Tres, Simonem Coliaeum, Paris, 1545, pp. 285, 287.

51 Colombo, Realdo, *De Re Anatomica*, Book XI, Nicholae Beuilacquae, Venice, 1559, pp. 262–69.

52 Falloppio, Gabriele, *Observationes Anatomicae*, 2 vols, Venice, 1561; reprinted S.T.E.M. Mucchi, Modena, 1964, p. 193.

53 Paul of Aegina, c.625–c.690, a 7th-century Byzantine Greek priest best known for his medical encyclopedia, published in seven books and referenced throughout the medieval and Renaissance periods.

54 Vesalius, Andreas, *Anatomicarum Gabrielis Falloppii Observationum Examen*, Francesco de Franceschi da Siena, Venice, 1564, p. 143.

55 De Graaf, Regnier, *Mulierum Organis Generationi Inservientibus Tractatus Novus Cum Figuris* ("Treatise on the Generative Organs of Women"), 1672.

56 Jocelyn, H. D. and B. P. Setchell (tr.), "Regnier de Graaf on the Human Reproductive Organs: An Annotated Translation of *De Mulierum Organis Generationi Inservientibus Tractatus Novus* (1672)," *Journal of Reproduction and Fertility Supplement*, No. 17, 1972, pp. 89–92.

57 Sharp, Jane, *The Midwives Book, or, the Whole Art of Midwifry Discovered: Directing childbearing women how to behave themselves*

in their conception, breeding, bearing, and nursing of children in six books, viz. ... / By Mrs. Jane Sharp practitioner in the art of midwifry above thirty years, printed for Simon Miller, at the Star at the West End of St. Paul's, London, 1671.

58 Morphis, Catherine (English department, University of Texas), "Swaddling England: How Jane Sharp's Midwives Book Shaped the Body of Early Modern Reproductive Tradition," *Early Modern Studies Journal,* Vol. 6, University of Texas, Arlington, English Department, 2014.

59 Lisa Forman Cody, "Introductory Note," *The Early Modern Englishwoman: A Facsimile Library of Essential Works, Printed Writings, 1641-1700,* Part 1, ed. Betty S. Travitsky and Patrick Cullen, Ashgate, Aldershot, 2002.

60 Sharp, *The Compleat Midwife's Companion,* Gale Ecco, 2010, pp. 43–46.

61 Lefkowitz Horowitz, Helen, *Rereading Sex: Battles over Sexual Knowledge and Suppression in Nineteenth-Century America,* Vintage, New York, 2003.

62 Full text of *Aristotle's Masterpiece* available at the Project Gutenberg eBook.

63 Jonathan Edwards manuscripts, Special Collections, Franklin Trask Library, Andover Newton Theological School, Newton, MA.

64 Carroll, Jason S; Padilla-Walker, Laura M.; Nelson, Larry J.; Olson, Chad D.; Barry, Carolyn McNamara; Madsen, Stephanie D., "Generation XXX: Pornography Acceptance and Use Among Emerging Adults," *Journal of Adolescent Research,* Vol. 23, 2008.

65 Kobelt, Georg L., *Die männlichen und weiblichen Wollustorgane des Menschen und einiger Säugetiere* ("The Male and Female Organs of Sexual Arousal in Man and Some Other Mammals"), 1884 (quoted in *The Classic Clitoris: Historic Contributions to Scientific Sexuality,* ed. T. P. Lowry, Nelson-Hall, Chicago, 1978, pp. 19–56).

66 O'Connell, Helen E.; K. V. Sanjeevan; J. M. Hutson, "Anatomy of the Clitoris," *Journal of Urology,* No. 174, 2005, pp. 1189–1195.

67 Darwin, Charles, *The Descent of Man and Selection in Relation to Sex,* Vol. 2, John Murray, London, 1871, p. 327.

68 Havelock Ellis, *Psychology of Sex,* 1913, p. 250.

69 Ibid., p. 236.

70 Duffy, Michael, "Getting Out the Wrecking Ball," *Time,* December 19, 1994.

71 Rodriguez, Sarah B., "Female Sexual Degeneracy and the Enlarged Clitoris, 1850–1941," *Female Circumcision and Clitoridectomy in the United States: A History of a Medical Treatment,* University of Rochester Press, Rochester, NY, 2014, pp. 49–74; Boydell & Brewer, Woodbridge, Suffolk, 2018, online publication.

72 Loudon, Irvine, "Maternal Mortality in the Past and Its Relevance to Developing Countries Today," *American Journal of Clinical Nutrition,* 2000, Vol. 72, Issue 1, pp. 241s–246s.

73 Baker, Smith, "The Neuro-Psychical Element in Conjugal Aversion," *Journal of Nervous and Mental Disease,* Vol. 17, No. 9, 1892, pp. 124–25.

74 For example, A. K. Gardner, "The Hygiene of the Sewing Machine," *American Medical Times* 1, 1860, and "Influence of Sewing Machine on Female Health," *New Orleans Medical and Surgical Journal* 20, November 1867.

75 Dickinson, Robert Latou, "Bicycling for Women from the Standpoint of the Gynecologist," *American Journal of Obstetrics and Diseases of Women and Children* 31, 1895, and W. E. Fitch, "Bicycle Riding: Its Moral Effect on Young Girls and Its Relation to Diseases of Women," *Georgia Journal of Medicine and Surgery* 4, 1899.

76 Maines, *Technology of Orgasm,* p. 160.

77 Havelock Ellis, *Psychology of Sex,* 1927, p. 176.

78 Kellogg, Dr. John Harvey, *Plain Facts for Old and Young, or the Science of Human Life from Infancy to Old Age,* Good Health Publishing Company, Battle Creek, MI, 1919, p. 325, www.babel. hathitrust.org. Earlier versions of this text were *Plain Facts about Sexual Life,* 1877, and *Plain Facts for Old and Young,* 1879. In 1886 it was 644 pages long; by 1901, 720 pages; by 1903, 798. In 1917 Kellogg published a four-volume edition of 900 pages. An

estimated half-million copies were sold, many by discreet door-to-door canvassers.

79 Kellogg, Dr. John Harvey, *Ladies' Guide in Health and Disease: Girlhood, Maidenhood, Wifehood, Motherhood,* Modern Medicine Publishing Co., Battle Creek, MI, 1893, pp. 550–551. www.books. google.com

80 A Gentleman of the Inner Temple, *A Digest of the Law Concerning Libels: Containing All the Resolutions in the Books on the Subject, and Many Manuscript Cases,* Libel and Slander, William Hallhead, Dublin, 1778, p. 60. www.books.google.com

81 Everett, L. S., pamphlet, *An Exposure of the Principles of the "Free Inquirers,"* Benjamin B. Mussey, Boston, 1831, pp. 34–35.

82 Landis, Simon, *A Full Account of the Trial of Simon M. Landis, M.D., for Uttering and Publishing a Book Entitled "Secrets of Generation,"* First Progressive Christian Church, Philadelphia, 1870 (quoted in Helen Lefkowitz Horowitz, *Rereading Sex,* see note 61).

83 Rabban, David, *Free Speech in Its Forgotten Years, 1870–1920,* Cambridge University Press, 1999, p. 39.

84 Orenstein, *Girls and Sex.*

85 SIECUS (Sexuality Information and Education Council of the United States), 2018.

86 Scoutetten, Dr. Henri, *Rapport sur l'Hydrotherapie,* V. Levrault, Strasbourg, 1843, pp. 239–41.

87 https://www.etymonline.com/word/frigidity

88 IMS Health, Retail and Provider Perspective, 1998, and National Prescription Audit, 1998.

89 Baker Brown, Isaac, *On the Curability of Certain Forms of Insanity, Epilepsy, Catalepsy and Hysteria in Females,* Robert Hardwick, London, 1866, p. 17.

90 Burt, James C., and Joan Burt, *Surgery of Love,* Carlton Press, New York, 1975 (quoted in John Griffith, *The Moral Challenges of Health Care Management,* Health Administrative Press, Ann Arbor, MI, 1993, p. 97).

91 Reported in *BMJ,* No. 456, 1866.

92 Webber, Sarah, and Toby Schonfeld, "Cutting History, Cutting

Culture: Female Circumcision in the United States," *The American Journal of Bioethics*, Vol. 3, No. 2, 2003, pp. 65–66.

93   *Merck's 1899 Manual of the Materia Medica*, Merck & Co., New York, 1899 (www.archive.org).

94   Morton, Mark, *The Lover's Tongue*, Insomniac Press, Toronto, 2003, p. 157.

95   Krafft-Ebing, Richard, *Psychopathia Sexualis: A Medico-Forensic Study*, G. P. Putnam's Sons, New York, 1896, p. 14.

96   Davis, Katharine B., *Factors in the Sex Life of Twenty-Two Hundred Women*, Harper, New York and London, 1929.

97   Havelock Ellis, *Psychology of Sex*, 1927, in his chapter, "The Sexual Impulse in Women," pp. 195–212.

98   Rodriguez, Sarah B., "Female Sexual Degeneracy," University of Rochester Press, pp. 49–74; Boydell & Brewer, p. 56.

99   I am indebted to Therese O'Neill's November 7, 2013, article "Advice for Your Wedding Night (from 100 Years Ago)" in *The Week* for the references that I use in this paragraph.

100 Stopes, Marie, *Married Love: A New Contribution to the Solution of Sex Difficulties*, 9th Ed., G. P. Putnam's Sons, London, 1920, p. 95. First published 1918 in U.K. and U.S.A.

101 Moore, Lisa Jean, and Adele E. Clarke, "Clitoral Conventions and Transgressions: Graphic Representations in Anatomy Texts, c. 1900–1991," *Feminist Studies*, Vol. 21, No. 2 (summer 1995), pp. 255–301.

102 Silverstein, Alvin, *Human Anatomy and Physiology*, John Wiley & Sons, New York, 1998, pp. 734, 740, 742.

103 Azadzoi, Kazem M., and Mike B. Siroky, "Neurologic Factors in Female Sexual Function and Dysfunction," *Korean Journal of Urology*, Vol. 51, No. 7, 2010, pp. 443–449.

104 McGovern, Kevin B.; Rita C. Stewart; Joseph Lo Piccolo, "Secondary Orgasmic Dysfunction. 1. Analysis and Strategies for Treatment," *Archives of Sexual Behavior*, Vol. 4, No. 3, 1975, pp. 265–275. Also, Schneidman, Barbara and Linda McGuire, "Group Therapy for Non-Orgasmic Women: Two Age Levels," *Archives of Sexual Behavior*, Vol. 5, No. 3, 1976, pp. 239–247.

105 Wallen, K., and E. Lloyd, "Female Arousal: Genital Anatomy and Orgasm in Intercourse," *Hormones and Behaviour,* 2011.

106 Oakley, S. H.; G. K. Mutema; C. C. Crisp; M. V. Estanol; S. D. Kleeman; A. N. Fellner; R. N. Pauls, "Innervation and Histology of the Clitoral-Urethal Complex: A Cross-Sectional Cadaver Study," *Journal of Sexual Medicine,* Vol. 10, 2013, pp. 2211–2218.

107 Gravina, G. L.; F. Brandetti; P. Martini; E. Carosa; S. M. Di Stasi; S. Morano; A. Lenzi; E. A. Jannini, "Measurement of the Thickness of the Urethrovaginal Space in Women with or without Vaginal Orgasm," *Journal of Sexual Medicine,* Vol. 5, 2008, pp. 610–618.

108 Komisaruk, Barry R; Beverly Whipple; Audrita Crawford; Sherry Grimes; Wen-Ching Liu; Andrew Kalnin; Kristine Mosier, "Brain Activation During Vaginocervical Self-stimulation and Orgasm in Women with Complete Spinal Cord Injury: fMRI Evidence of Mediation by the Vagus Nerves," *Brain Research,* No. 1024, 2004.

109 Foldès, Pierre, and Odile Buisson, "The Clitoral Complex: A Dynamic Sonographic Study," *The Journal of Sexual Medicine,* Vol. 6, No. 5, May 2009, pp. 1223–31.

110 Whipple, Beverly, "Female Ejaculation, G Spot, A Spot, and Should We Be Looking for Spots?" *Current Sexual Health Report,* No. 7, 2015, pp. 59–62.

111 Foldès, Pierre, M.D.; Dr Béatrice Cuzin, M.D.; Armelle Andro, PhD, "Reconstructive Surgery after Female Genital Mutilation: A Prospective Cohort Study," *The Lancet,* Vol. 380, No. 9837, July 2012, pp. 134–141.

Part Two: Perception

112 Philo, 13 BCE–54 CE, *Questions and Answers on Genesis,* tr. Ralph Marcus, from the ancient Armenian version of the original Greek, Harvard University Press, Cambridge, MA, 1953.

113 Clark, Russell D., and Elaine Hatfield, "Gender Differences in

Receptivity to Sexual Offers," *Journal of Psychology and Human Sexuality*, Vol. 2, 1989, pp. 39–55.

114 Henderson, Katherine Usher, and Barbara F. McManus, *Half Humankind: Contexts and Texts of the Controversy about Women in England, 1540–1640*, University of Illinois Press, Champaign, IL, 1985.

115 Tattlewell, Mary, and Joan-Hit-Him-Home, *The Women's Sharp Revenge*, London, 1640, pp. 133–134. The pen names are indicative of the banteresque nature of the pamphlet wars.

116 *Cosmopolitan* survey, 2015.

117 This statistic comes from not just one study, but a comprehensive analysis of 35 studies over the past 80 years by Elisabeth Lloyd in *The Case of the Female Orgasm*.

118 Meehan, Ciara, "Has He Called You Frigid Lately?" *History Ireland*, Vol. 26, No. 4, 2018, pp. 44–46. www.jstor.org/stable/26565900

119 Orenstein, *Girls and Sex*, p. 89.

120 Henry Kaiser Family Foundation, 2003, and Advocates for Youth, 2002 (quoted in Orenstein, *Girls and Sex*, p. 246).

121 CCC, No. 1607, alluding to Genesis 1:28, 2:22, 3:12, and 3:16–19, quoted in *Marriage: Love and Life in the Divine Plan: A Pastoral Letter of the United States Conference of Catholic Bishops*, USCCB, November 17, 2009, pp. 19, 29. © 2009, United States Conference of Catholic Bishops.

122 Pope Francis, *Amoris Laetitia*, Ch. 4, "Love in Marriage," 2016, p. 152.

123 McCann, Christine A., "Transgressing the Boundaries of Holiness: Sexual Deviance in the Early Medieval Penitential Handbooks of Ireland, England and France 500–1000," *Theses* 76, Seton Hall University, NJ, 2010. www.scholarship.shu.edu/theses/76

124 Letter to Nicolas Gerbel. See also James Reston Jr., *Luther's Fortress: Martin Luther and His Reformation Under Siege*, Basic Books, New York, 2015.

125 Hesiod, *Works and Days*, 700 BCE. www.thoi.com/text/HesiodTheogony

126 Aristotle, "On the Nature of Women," *Historia animalium* ("History of Animals"), Book IX, 350 BCE; one can find similar statements in his other famous text, *Politics*.

127 Lowen, Alexander, *Love and Orgasm*, Macmillan, New York, 1965, p. 216.

128 Debay, Auguste, *Hygiène et Physiologie du Mariage*, Moquet, Paris, 1848.

129 Alexander, Stephanie, "Was It Good For You Too?" *Cosmopolitan*, No. 5, 1995, p. 80.

130 Bible, book of Revelation, Ch. 17.

131 www.donsmaps.com/vulvastoneage.html is a comprehensive site that lists examples in an accessible format.

132 Guthrie, R. Dale, *The Nature of Paleolithic Art,* University of Chicago Press, 2005.

133 Gilbert, Harriet, and Christine Roche, *A Women's History of Sex*, Pandora Press, London, 1987.

134 Philo, *Questions and Answers on Genesis,* 13 BCE–54 CE.

135 https://plato.stanford.edu

136 Freud, Sigmund, "The Sexual Life of Human Beings," *Introductory Lectures on Psycho-Analysis*, Lecture XX, W.W. Norton, New York, and Liveright, New York, 1966, p. 394; originally published 1917.

137 "For women the level of what is ethically normal is different from what it is in men. Their superego is never so inexorable, so impersonal, so independent in its emotional origins as we require it to be in men. Characteristic traits which critics of every epoch have brought up against women—that they show less sense of justice than men, that they are less ready to submit to the great exigencies of life, that they are more often influenced in their judgements by feelings of affection or hostility—all these would be amply accounted for by the modification in the formation of their superego which we have inferred above." Freud, in "Some Psychological Consequences of the Anatomical Distinction Between the Sexes," 1925. www.aquestionofexistence.com

138 Freud, Sigmund, "Female Sexuality," *Introductory Lectures*, 1931,

and "Femininity," 1933, *New Introductory Lectures on Psycho-Analysis*, W.W. Norton, New York, 1990.

139 Ibid., "Some Psychological Consequences," 1933.

140 Ibid., "The Sexual Life of Human Beings," *Introductory Lectures*, 1917.

141 Ibid., 1940. I have been unable to find the primary source for this quotation. If you have it, I'd love it. Please DM me on Instagram via @its.personalgirls.

142 Ibid., "Femininity," 1933.

143 Kaplan, Helen Singer, *The New Sex Therapy: Active Treatment of Sexual Dysfunctions*, Bailliere Tindall, London, 1974, p. 544

144 Kardiner, Abram, review of Kinsey's *Sexual Behavior in the Human Female* for *The American Scholar*, Vol. 23, No. 1 (winter 1953/54), pp. 106, 108, 110.

145 Questions that refer to the clitoris in any way at all in *The Hite Report*: In "Orgasm" section—Q.3 Do you have orgasms during the following (please indicate whether always, usually, sometimes, rarely, or never): masturbation/intercourse (vaginal penetration)/manual clitoral stimulation by a partner/oral stimulation by a partner/intercourse plus manual clitoral stimulation/never have orgasms. Also indicate above how many orgasms you usually have during each activity, and how long you usually take. In "Sexual Activities" section—Q.15 Do(es) your partner(s) stimulate your clitoral area manually? How? Is it usually for purposes of orgasm or arousal? If for orgasm, does it lead to orgasm always, usually, sometimes, rarely, or never? Is this form of sex important to you? Q.16 Do(es) your partner(s) stimulate you orally (cunnilingus)? Is this stimulation oral/clitoral or oral/vaginal, or both? Is it for orgasm or arousal? If for orgasm, does it lead to orgasm always, usually, sometimes, rarely, or never? Do you like it? Q.19 If you orgasm during vaginal penetration/intercourse, are other accompanying stimuli usually present? What would you say is your method of obtaining clitoral stimulation during intercourse: a) long foreplay, b) simultaneous manual stimulation of the clitoris, c) indirect stimulation from thrusting, d) "grinding" or pressing together during penetration, or e) some other method? Q.23 Is

it easier for you to have an orgasm by clitoral stimulation when intercourse is not in progress? If you had to choose between intercourse and clitoral stimulation by your partner, which would you pick? Why?

146 Bancroft, John, "Sexual Science in the 21st Century: Where Are We Going? A Personal Note," *The Journal of Sex Research*, Vol. 1, No. 36, 1999, pp. 226–229.

147 Foley, Katherine Ellen; Youyou Zhou; Christopher Groskopf, "Our Analysis of Five Decades of Sex Research Shows an Evolving Spectrum of Sexual Norms," Quartz, September 21, 2017. www.qz.com

148 Aristotle, *Historia animalium*, 10.5.637a, pp. 23–25.

149 Laqueur, *Making Sex*, p. 37.

150 Fox, C. A., and B. Fox, "A Comparative Study of Coital Physiology with Special Reference to the Sexual Climax," *Journal of Reproduction and Fertility*, Vol. 24, 1971, pp. 319–336.

151 Wildt, L.; S. Kissler; P. Licht; W. Becker, "Sperm Transport in the Human Female Genital Tract and Its Modulation by Oxytocin as Assessed by Hysterosalpingoscintigraphy, Hysterotonography, Electrogysterography and Doppler Sonography," *Human Reproduction Update*, Vol. 4, No. 5, 1998, pp. 655–666.

152 Elliot, Mark (director of the Institute for Psychological and Sexual Health, Columbus, OH), "Checking the Male," *Men's Health*, 2005.

153 Hewlings, Susan J., and Douglas S. Kalman, "Curcumin: A Review of Its Effects on Human Health," *Foods*, Vol. 6, No. 10 (October 22, 2017), p. 92, Basel, Switzerland. doi:10.3390/foods6100092

154 Levin, Roy H., "Can the Controversy About the Putative Role of the Human Female Orgasm in Sperm Transport be Settled with Our Current Physiological Knowledge of Coitus?" *The Journal of Sexual Medicine*, Vol. 8, No. 6 (June 1, 2011), pp. 1566–1578.

155 Bernds, W.P, and D. Barash, "Early Termination of Parental Investment in Mammals, including Humans," *Evolutionary Biology and Human Social Behaviour: An Anthropological*

Perspective, ed. N. Chagnon and W. Irons, Duxbury Press, North Scituate, MA, 1979, pp. 487–505.

156 Morris, Desmond, *The Naked Ape: A Zoologist's Study of the Human Animal*, McGraw-Hill, New York, 1967, p. 65.

157 "An Evolutionary Whodunit: How Did Humans Develop Lactose Tolerance?" interviewed by Helen Thompson, December 2012.

158 Sherfey, Mary Jane, *The Nature and Evolution of Female Sexuality*, Random House, New York, 1973.

159 In the Kinsey sample (1953) it was found that 62 percent of women masturbated; in Hite's sample (1976), 82 percent; the *Journal of Sex Research* survey of 2016 found that 85.5 percent of women in their sample masturbated.

160 Lloyd, *Female Orgasm*, p. 111.

161 Symons, Donald, *The Evolution of Human Sexuality*, Oxford University Press, New York, 1979. See also "Precis of the Evolution of Human Sexuality," *Behavioral and Brain Sciences*, Vol. 3, 1980, pp. 171-181.

162 Thornhill, Randy, and Steven W. Gangestad, *The Evolutionary Biology of Human Female Sexuality*, Oxford University Press, 2008; ProQuest Ebook Central, pp. 29–30.

Part Three: Representation

163 Colombo, Realdo, "amor Veneris, vel dulcedo appelletur," *De Re Anatomica*, 1555.

164 Ogletree, Shirley, and Harvey Ginsberg, the PsycINFO database.

165 National Coalition for Men (NCFM), *174 Ways to Call a Penis Something Other Than "Penis"!* San Diego, 8 June 2011. "It appears that the word 'Penis' may have more synonyms than any other word in the English language."

166 www.glamourmagazine.co.uk/gallery/vulva-vagina-facts-you-didnt-know

167 Cocking, Lauren, "On the Gleefully Indecent Poems of a Medieval Welsh Feminist Poet," Literary Hub, August 9, 2019. www.lithub.com

168 https://en.wiktionary.org/wiki/Thesaurus

169 https://onlinelibrary.wiley.com

170 Rowling, J. K., *Harry Potter and the Philosopher's Stone*, Ch. 17, "The Man with Two Faces," 1997.

171 Frank, Anne, *The Diary of a Young Girl: The Definitive Edition*, ed. Otto H. Frank and Mirjam Pressler, Puffin, London, 1997.

172 https://www.penguinrandomhouse.com/authors/8834/gustave-flaubert

173 Flaubert, Gustave, *Madame Bovary*, Barnes & Noble Classics, New York, 2005, p. 150; originally published 1857.

174 Rousseau, Jean-Jacques, *"Emile, or On Education,"* Paris, 1772, and London, 1773.

175 Cleland, John, *Memoirs of a Woman of Pleasure (Fanny Hill)*, 1748. www.googlebooks.com

176 Ibid.

177 Aphra Benn, 1640–1689; Mary Wollstonecraft, 1759–1797; Jane Austen, 1775–1817; George Sand (real name, Amantine Lucile Aurore Dupin), 1804–1876; Emily Brontë (pen name, Ellis Bell), 1816–1855; George Eliot (real name, Mary Ann Evans), 1819–1880.

178 "Teens and Porn: 10 Stats You Need Know," Covenant Eyes, August 19, 2010. www.covenanteyes.com/2010/08/19/teens-and-porn-10-stats-your-need-to-know

179 Orenstein, *Girls and Sex*.

180 Carroll, Jason E., et al, "Generation XXX," pp. 6–30.

181 Orenstein, *Girls and Sex*, pp. 32–38.

182 Silverberg, Cory, and Fiona Smyth, *Sex Is a Funny Word: A Book about Bodies, Feelings, and You*, Triangle Square, New York, 2015.

183 Gravelle, Karen, *The Period Book: A Girl's Guide to Growing Up*, Bloomsbury, New York, 2017, pp. 31–32.

184 https://www.plannedparenthood.org/about-us/newsroom/press-releases/new-poll-parents-talking-their-kids-about-sex-often-not-tackling-harder-issues

185 http://www.advocatesforyouth.org/parents

186 ReCAPP Resource Center for Adolescent Pregnancy Prevention.

187 Orenstein, *Girls and Sex*.

188 Kirby, D., *Emerging Answers 2007: Research Findings on Programs to Reduce Teen Pregnancy and Sexually Transmitted Diseases*, National Campaign to Prevent Teen and Unplanned Pregnancy, Washington, D.C., 2007.

189 Gravelle, Karen, *What's Going On Down There? A Boy's Guide to Growing Up,* Bloomsbury, New York, 2017.

190 "The Pleasure is Ours," *Gloop Lab,* Episode 3, Netflix, 2020.

191 The Archibald Fountain, with Theseus slaying the Minotaur, is outside St. James Church, Hyde Park, Sydney. As if he didn't learn his lesson there, Theseus can also be found slaying the Minotaur naked in the Jardin des Tuileries, Paris.

192 Plutarch (c. 46–120 CE), "The Persian Women," *The Bravery of Women,* section V.

193 Kristeva, Julia, "Powers of Horror: An Essay on Abjection," tr. Leon S. Roudiez, European Perspectives, Columbia University Press, New York, 1982, pp. 77, 100.

194 Weir, Anthony, and James Jerman, *Images of Lust: Sexual Carvings on Medieval Churches,* Routledge, London, 1986.

195 O'Donovan, John, in his Ordnance Survey letters, 1840.

196 Smith, Lewis, "'Notorious heterosexual' Lucian Freud's letters suggest he had an affair with Stephen Spender," *Independent,* June 7, 2015.

197 http://www.artic.edu/aic/collections/artwork/24687

198 thepornconversation.org is a website designed specifically to help parents talk to children about porn.

199 Bridges A.J.; R. Wosnitzer; E. Scharrer; C. Sun; R. Liberman, "Aggression and Sexual Behaviour in Best-Selling Pornography Videos," Sage Journals, October 26, 2010.

200 Reported by Ammar Ebrahim, BBC Stories, April 5, 2019.

Part Four: Climax

201 Hurlbert, David Farley, and Karen Elizabeth Whittaker, "The Role of Masturbation in Marital and Sexual Satisfaction: A Comparative Study of Female Masturbators and Non Masturbators," *Journal of Sex Education and Therapy,* Vol. 17, No. 4, 1991, pp. 272–282.

202 Baumeister, R. F.; K. R. Catanese; K. D. Vohs, "Is There a Gender Difference in Strength of Sex Drive? Theoretical Views, Conceptual Distinctions, and a Review of Relevant Evidence," *Personality and Social Psychology Review,* Vol. 5, No. 20, pp. 242–273.

203 Fisher, T.; Z. Moore; M. Pittenger, "Sex on the Brain? An Examination of Frequency of Sexual Cognitions as a Function of Gender, Erotophilia, and Social Desirability," *The Journal of Sex Research*, Vol. 49, No. 1, 2011, pp. 69–77.

204 Armstrong, Elizabeth A.; Paula England; Alison C. K. Fogarty, "Accounting for Women's Orgasm and Sexual Enjoyment in College Hookups and Relationships," *American Sociological Review*, May 7, 2012.

205 Conley, Terri D.; Amy C. Moors; Jes L. Matsick; Ali Ziegler; Brandon A. Valentine, "Women, Men, and the Bedroom: Methodological and Conceptual Insights that Narrow, Reframe, and Eliminate Gender Differences in Sexuality," *Current Directions in Psychological Science*, Vol. 20, No. 5, October 2011, pp. 296–300; Sage Publications on behalf of Association for Psychological Science. https://www.jstor.org/stable/23045742

206 González-Alvarez, Julio, and Teresa Cervera-Crespo, "Gender Differences in Sexual Attraction and Moral Judgment: Research with Artificial Face Models," *Psychological Reports*, Vol. 122, No. 2, 2019, pp. 525–535.

207 Whipple, "Female Ejaculation," p. 61.

Appendix: FGM

208 *Bolokoli*—meaning "washing your hands," Bambara, used in Mali. Zabus, Chantal, "From 'Cutting Without Ritual' to 'Ritual Without Cutting': Voicing and Remembering the Excised Body in African Texts and Contexts," Sorbonne, 2008, (in Borch, Merete Falck, ed., *Bodies and Voices: The Force-Field of Representation and Discourse in Colonial and Postcolonial Studies*, Rodopi, New York, 2007). *Bondo*—an initiation rite into adulthood in Sierra Leone. Schweder, Richard A., "Disputing the Myth of the Sexual Dysfunction of Circumcised Women," an interview with Fuambai S. Ahmadu, *Anthropology Today*, Vol. 25, No. 6, December 2009. *Silidjili*—meaning ablution or ritual purity, French/Bambara, used in Mali. Safeguarding Children Board and National FGM Center, Milton Keynes, U.K. *Thara*—Egyptian, deriving from the Arabic *tahur/tahara*

meaning to clean/purify; known as *thoor* in Sudan. El Guindi, Fadwa, "Had This Been Your Face, Would You Leave It as Is?" 2007. Abusharaf, Rogaia Mustafa (ed.), *Female Circumcision: Multicultural Perspectives,* University of Pennsylvania Press, Philadelphia, p. 30.

209 World Health Organization fact sheet, February 3, 2020.

210 NHS Enhanced Dataset, collected by NHS Digital, April 2015 to March 2020.

211 https://www.unfpa.org/resources/female-genital-mutilation-fgm-frequently-asked-questions

212 Data from the World Bank, www.datatopics.worldbank.org/world-development-indicators/stories/fgm-still-practiced-around-the-world.html

213 Ibid.

214 UNIFPA https://sustainabledevelopment.un.org/content/documents/19961027123_UN_Demograhics_v3%20(1).pdf

215 Batha, Emma, "Factbox: Female Genital Mutilation around the World: A Fine, Jail or No Crime?" Reuters, September 13, 2018.

216 Schweder, "Disputing the Myth," interview with Fuambai S. Ahmadu.

217 Goldstein, Ruth, "Talking Drums and Ethical Conundrums," *Anthropology Matters Journal,* Vol. 12, No. 1, 2010, p. 10.

Select Bibliography

Abusharaf, Rogaia Mustafa (ed.), *Female Circumcision: Multicultural Perspectives*, University of Pennsylvania Press, Philadelphia, 2007.

Alberti, Fay Bound, *This Mortal Coil: The Human Body in History and Culture*, Oxford University Press, on behalf of the Society for the Social History of Medicine, 2016.

Anolik, Ruth Bienstock, *Horrifying Sex: Essays on Sexual Difference in Gothic Literature*, McFarland & Company, Jefferson, NC, 2007.

Armstrong, David, "Bodies of Knowledge: Foucault and the Problem of Human Anatomy," *Psychology*, 2003.

Bonnard, Jean-Baptiste, "Male and Female Bodies According to Ancient Greek Physicians," tr. Lillian E. Doherty and Violaine Sebillotte Cuchet, *Clio*, Vol. 7, January 2013.

Brundage, James A., *Law, Sex, and Christian Society in Medieval Europe*, University of Chicago Press, 1987.

Cott, Nancy F., "Passionlessness: An Interpretation of Victorian Sexual Ideology, 1790–1850," *Signs*, Vol. 4, No. 2, 1978, pp. 219–236.

Criado-Perez, Caroline, *Invisible Women: Exposing Data Bias in a World Designed by Men*, Abrams Press, New York, 2019.

Cryle, Peter, and Alison Moore, *Frigidity: An Intellectual History*, Palgrave Macmillan, Basingstoke, 2011.

Dijkstra, Bram, *Idols of Perversity: Fantasies of Feminine Evil in Fin-de-Siècle Culture*, Oxford University Press, New York, 1986.

Dijkstra, Bram, *Evil Sisters: The Threat of Female Sexuality and the Cult of Manhood*, Alfred A. Knopf, New York, 1996.

Ehrenreich, Barbara, and Deirdre English, *For Her Own Good: Two Centuries of the Experts' Advice to Women*, Anchor Books, New York, and Random House, Canada, 1978 and 2005.

Goldstein, Ruth, "Talking Drums and Ethical Conundrums," *Anthropology Matters Journal*, Vol. 12, No. 1, 2010.

Haynes, April R., *Riotous Flesh: Women, Physiology, and the Solitary Vice in Nineteenth-Century America*, The University of Chicago Press, 2015.

Hite, Shere, *The Hite Report: A Nationwide Study on Female Sexuality*, Macmillan, New York, 1976.

Hite, Shere, *The Hite Report: Women and Love: A Cultural Revolution in Progress*, Knopf, New York, 1987.

Horowitz, Helen Lefkowitz, *Rereading Sex: Battles over Sexual Knowledge and Suppression in Nineteenth-Century America*, Vintage Books, New York, 2003.

Kinsey, Alfred, *Sexual Behavior in the Human Female*, W. B. Saunders Company, Philadelphia and London, 1953.

Laqueur, Thomas, *Making Sex: Body and Gender from the Greeks to Freud*, Harvard University Press, Cambridge, MA, 1990.

Lloyd, Elisabeth A., *The Case of the Female Orgasm: Bias in the Science of Evolution*, Harvard University Press, Cambridge, MA, 2005.

Maines, Rachel P., *The Technology of Orgasm: "Hysteria," the Vibrator, and Women's Sexual Satisfaction*, Johns Hopkins University Press, Baltimore, MD, 2001.

Masters, William H., and Virginia E. Johnson, *Human Sexual Response*, Bantam Books, New York and Toronto, 1966.

Masters, William H. and Virginia E. Johnson, *Human Sexual Inadequacy*, Bantam Books, New York and Toronto, 1970.

Matus, Jill L., *Unstable Bodies: Victorian Representations of Sexuality and Maternity*, Manchester University Press, 1995.

Odem, Mary E., *Delinquent Daughters: Protecting and Policing Adolescent Female Sexuality in the United States, 1885–1920*, North Carolina Press, Chapel Hill, NC, 1995.

Orenstein, Peggy, *Girls and Sex: Navigating the Complicated New Landscape,* Harper, New York, 2016.

Orenstein, Peggy, *Boys and Sex: Young Men on Hook-Ups, Love, Porn, Consent and Navigating the New Masculinity,* Souvenir Press, London, 2020.

Rodriguez, Sarah B., *Female Circumcision and Clitoridectomy in the United States: A History of a Medical Treatment,* University of Rochester Press, New York, 2014.

Scully, Diana, and Pauline Bart, "A Funny Thing Happened on the Way to the Orifice: Women in Gynecology Textbooks," *American Journal of Sociology,* Vol. 78, No. 4, January 1973.

Showalter, Elaine, *The Female Malady: Women, Madness, and English Culture, 1830–1980,* Virago, London, 1987.

Walkowitz, Judith R., *Prostitution and Victorian Society: Women, Class, and the State,* Cambridge University Press, 1980.

Walkowitz, Judith R., *City of Dreadful Delight: Narratives of Sexual Danger in Late Victorian London,* University of Chicago Press, 1992.

Picture Credits

All cartoons: author's own.

Cover

Design: Eilidh Doig Design, www.eilidhdoig.myportfolio. com, for the author.

Venus: from a painting by Henri Pierre Picou, 1824–1895.

Microscope: iStock/Nastasic.

Hand: iStock/Nicoolay.

Part One: Anatomy

Opening illustration: inspired by Jacopo Berengario da Carpi's anatomy text, *Isagoge breves*, 1522; by Eilidh Doig, for the author.

Female anatomical drawings, pp. 15, 16: Andreas Vesalius, *De Humani Corporis Fabrica Libri Septem*, detail from Book 5, The Organs of Nutrition and Generation, 1543.

Pelvic douche illustration, p. 29: from Fleury, c. 1860, reproduced from Sigfried Giedion, *Mechanization Takes Command*, Oxford University Press, New York, 1948.

Clitoris and vestibular bulbs illustration, p. 45: Georg Ludwig Kobelt, *Die männlichen und weiblichen Wollust-Organe des Menschen und einiger Säugethiere: in anatomisch-physiolo, Beziehung*, Freiburg i.Br., 1844 Tefel_3.

3D-printed clitoris, p. 73: photograph courtesy of Dr Christina Weis, Research Fellow, Centre for Reproduction Research, De Montfort University.

Female genital anatomy line drawings, pp. 71, 72: author's own.

Clitoral anatomy drawings, pp. 81: Hermione Chadwick.

Vulvarities, p. 82: courtesy of Katja Tetzlaff, medical illustrator specializing in diverse representations. www. ktetzlaff.com/IG: @ktetzlaffstudio

Part Two: Perception

Opening illustration: inspired by Auguste Rodin's *The Thinker*; by Eilidh Doig, for the author.

Carved vulva, p. 108: photograph by Don Hitchcock, 2008. www.donsmaps.com

Venus of Monpazier, p. 109: photograph courtesy of Professor Randall White (retired), Center for the Study of Human Origins, NYU. https://wp.nyu.edu/csho/people/ faculty/randall_white

Part Three: Representation

Opening illustration: inspired by Antonio Canova's *The Three Graces*; by Eilidh Doig, for the author.

Porta Tosa, Milan, p.193: photograph by G. Dallorto, own work, 2007. https://commons.wikimedia.org

Cerne Abbas giant, p. 197: photograph by Pete Harlow, own work, 2001. CC BY-SA 3.0, https://commons.wikimedia.org

Womanspreading placard, p. 203: by kind permission of Rumisa Lakhani and Rashida Muqadam. @rumisalakhani. art on Facebook and IG.

Part Four: Climax

Opening illustration: by Eilidh Doig, for the author.

Author photograph: taken by a lovely friend.

About the Author

Sarah Chadwick began writing this book after a family move from the U.K. to Chicago in 2016. "I had always wanted to write, but once I had the idea for *The Sweetness of Venus* it was as if my career came together—my love of reading, teaching, research, studying, libraries, high and low culture, and writing all seemed to coalesce with this project."

Sarah studied at Durham University, Kings College London, and Warwick University. She has four children and travels between Cornwall and Chicago. She also runs the @its.personalgirls Instagram page and can be found on facebook.com/sarah.chadwick.author.

CPSIA information can be obtained
at www.ICGtesting.com
Printed in the USA
BVHW031356260221
PP11915700002B/5